The  of
Bee Improvement

Jo Widdicombe

The Principles of Bee Improvement

Published in the United Kingdom by
Northern Bee Books,
Scout Bottom Farm,
Mytholmroyd,
West Yorkshire HX7 5JS
Tel: 01422 882751
Fax: 01422 886157

www.northernbeebooks.co.uk

ISBN 978-1-908904-62-1

Edited by John Phipps
Design and artwork, D&P Design and Print
Printed by Lightning Source (UK)

The Principles of
Bee Improvement

Jo Widdicombe

PREFACE

BIBBA (The Bee Improvement and Bee Breeders' Association) has led the way in bee improvement in Britain and Ireland. I have been involved in a local bee improvement group (BIPCo, the Bee Improvement Programme for Cornwall) one of many groups set up in order to put BIBBA's ideas into practice. To some extent we, in BIPCo, have regarded the process of bee improvement by small-scale beekeepers as an experiment to see if progress can, in fact, be made by beekeepers operating on a relatively small scale. The aim is to show that improvement in our stock can be achieved, and that these methods offer a real alternative to the common belief that improvement can only be made by bringing in new stock. The system of regularly bringing in new queens allows for no local adaptation and no thought is given as to how to maintain any improved quality, other than by repeating the process and importing further new stock.

Beekeepers are often tempted by the use of imported stock because of the prolific nature of the queens and the docility of the colonies produced. Other attributes, particularly their hardiness and ability to perform in our conditions are untested, and their docility is often quickly lost in succeeding generations. Further imports are then thought to be necessary in order to raise the standards again.

If more beekeepers choose the alternative route, of trying to 'improve what we have got' rather than that of constant importation, the long-term prospects for beekeeping in this country will be brighter. Instead of the endless battle against the decline in quality of our bees and the experience of high colony losses, we could witness a steady improvement in the quality of our stock. A more uniform, but genetically diverse, bee could be produced using locally-reared queens with qualities of hardiness, ease of management, productivity and good-temper.

This book deliberately offers a 'low-tech' approach, relevant to, and achievable by, all beekeepers.

Jo Widdicombe
January 2015

BIOSECURITY: A HEALTH WARNING

The process of 'Bee Improvement' can be a high risk strategy with regards to bee health. The process will involve the use of numerous colonies, often belonging to different beekeepers, the transfer and mixing of frames and brood when making up nucs, and the movement of colonies between apiaries, perhaps for the purposes of queen mating. All these processes increase the risk of the introduction of disease and once in the system can lead to their spread, far and wide, as nucs are distributed.

This would not only lead to the loss of reputation and trust in the quality of stock produced but could also put the whole improvement programme into jeopardy for years to come.

In particular, the notifiable diseases of American Foul Brood and European Foul Brood must be guarded against. It is recommended that all beekeepers involved in a bee improvement programme have their stock regularly checked by an approved bee inspector. This becomes even more important when any participating beekeeper is operating in an area with a history of foul brood.

By adopting a sensible approach to bio-security from the outset, serious problems can be avoided later on.

CONTENTS

PART 1:

The Theory of Bee Improvement   5

PART 2:

The Practical Aspects of Bee Improvement   23
1. Assessment of Stock (i) Behavioural characteristics   23
2. Assessment of Stock (ii) Selecting within a strain   30
3. Record Keeping   43
4. Queen rearing methods   45
5. Queen mating – The use of nucs and mini-nucs   55
6. Queen mating – The role and importance of the drone   62
7. Working as a group   67

PART 3:

Summary of 'The Principles of Bee Improvement'   69

CONCLUSION   73

APPENDIX (i)   74

APPENDIX (ii)   75

REFERENCES   76

ACKNOWLEDGEMENTS   77

INDEX   78

PART 1:

THE THEORY OF BEE IMPROVEMENT

Why bother with bee improvement?

The benefits of beekeeping extend well beyond that of honey production; the value in increased yields of agricultural crops, due to pollination by honey bees, greatly outweighs that of the honey produced by the beekeeper. In the wider environment, pollination increases 'seed set' benefitting the natural flora and fauna and maintaining biodiversity. Direct productivity, in the form of honey production can, therefore, be seen as a secondary issue.

Since the advent of *Varroa* and the demise of the feral population, honey bees have become more dependent on the beekeeper than ever before. Beekeeping has a history of being a pastime of the 'well-off' but has also provided opportunities for some financial return. A hundred years ago, it was seen as a pursuit which could provide useful extra income 'for the labouring classes'.

Currently there is a view that the importance of the honey bee is so great that we should not concern ourselves with the direct productivity of the bee, namely honey production. We should just support the bee for the agricultural and environmental benefits that it brings. Indeed it is argued that the exploitation of the bee for the production of honey is having a negative effect on the bee's welfare and is, therefore, counter-productive.

Like in any other food producing system, the aim should be for sustainability, and if we cannot produce honey without the demise of the honey bee or other negative or unsustainable effects on the environment, then we are doing something wrong.

At the present time, in many cases, we are in the historically unusual position of not needing to squeeze maximum food production out of the land. Similarly, in beekeeping, we can keep bees for our own pleasure or for the benefit that they bestow on the environment and not worry whether they are productive or not. However, the current position may be a luxury that will change over time. As world population grows, food production will increase in importance and honey production will play a role in that.

In the UK, with our dense population and our fickle climate, we produce about a sixth of the honey that we consume. As with many other foodstuffs, we are never likely to become self-sufficient in honey production, but the amount produced is economically significant and likely to increase in importance in the future.

Bees like any other organism vary greatly in their characteristics and the behaviour of one colony can be quite different to that of another. Most beekeepers, not just beginners, prefer a good tempered bee; beekeeping is more pleasurable, and the risks to the general public are less.

Beekeeping is enjoying a resurgence at the present time, to a large extent due to concerns over the viability or long-term survival of the honey bee. High losses of colonies cause alarm and although the numbers of beekeepers are increasing, little thought is given to how we maintain a sustainable bee population.

'Bee improvement' is relevant to beekeeping for several reasons. Bees, as an organism, need to be sustainable, that is hardy and able to survive in our environment; they need to be good tempered and not aggressive, so that they can happily co-exist with us on our densely populated island; they need to be productive so that our bee population provides a return for the beekeeper and is therefore economically sustainable.

Bee improvement – a better approach

Bee improvement is the process of maintaining and improving the quality of our bees. This has largely been regarded as something best left to commercial bee breeders, the common belief being that ordinary beekeepers merely need to buy in new queens from time to time in order to maintain or raise standards. According to this school of thought, it is advocated that bringing in new queens be repeated whenever the inevitable decline in quality is experienced.

We have been importing queens into the UK since 1859, originally for the purposes of bee improvement. Foreign queens have been billed as more docile, more prolific and therefore more productive, and this, combined with the novelty factor and the feeling that we can buy better stock than what we already have, has produced a good market for imported queens.

The importation of queens has become big business and nowadays most imports are not to 'improve' our bees but to satisfy the demand for new colonies. Imports offer a cheap source of queens which are used to produce colonies to supply new beekeepers as well as to replace natural losses. The ready availability of imported queens and the docility of colonies produced make them particularly attractive to new beekeepers as does the fascination of experimenting with different breeds of bee.

These imported queens can be impressive for their prolificacy but the large colonies produced can make management difficult. A very large colony may result in excellent yields of honey during times of good weather and reliable nectar flows. However in spells of changeable weather, which our climate frequently offers, heavy consumption of stores leads to a loss of productivity and the threat of starvation often results. The large colonies produced mean that inspections may require more time and effort, and certainly more skill in the handling of large numbers of bees, as well as a more 'hands-on' approach to management during adverse conditions.

The winter-hardiness of imported stock may be an issue and perhaps accounts for the high losses of stock amongst beginners. A high rate of loss, however, merely stimulates the demand for further imports in the following season. A decline in the temper of bees over succeeding generations also stimulates demand as, once the docility is lost, the beekeeper looks to buy better tempered stock.

After swarming, the new queen mates with local drones and the character of the colony may change for the worse, sometimes after 2 or 3 generations. A bad tempered colony is often blamed on the 'bad' characteristics of the native bee which are introduced into the genetic mix by the local drones. In reality, the bad temper and other unreliable qualities are the end result of the hybridisation of the different sub-species of honey bees. When the temper deteriorates, more imports are called for, causing further hybridisation and the setting up of a vicious circle of spiralling decline in quality.

Apart from the issues of temper, the suitability of these imported bees to perform in our conditions is not taken into account. By choosing imported stock, we are committing ourselves to a bee which has been selected to thrive in more productive environments than our own. Their performance in our more marginal honey producing conditions is a factor which has not been considered. We should be aiming to produce a bee which is more reliable and productive in our particular circumstances, rather than using ones designed for performance in areas with long spells of good weather and reliable nectar flows, as found in the major honey producing countries of the world.

Importation may increase the health risks faced by our bees. It is known that different strains of American and European Foul Brood exist around the world. Importation not only exposes our bees to the risk of bringing in new pathogens, but perhaps, more likely, there is the danger of bringing in different, and possibly more virulent, strains of the diseases that our bees are already exposed to. Let us not forget that the Isle of Wight disease was probably a pathogen imported into this country on another sub-species of bee.

What effect is this large-scale importation, of around 10,000 queens per year, having on our bee population? One thing most beekeepers will agree on is that the quality of our bees is not high. They are very variable in their performance and temper and, because of the genetic mix, it is difficult to produce any improved reliability in succeeding generations. The demand for a better bee goes on, encouraging keen beekeepers to continue to look abroad for more consistent stock and to buy new queens every time a decline in quality is experienced. This will continue to be the case until a viable alternative is offered.

It is time to ask the question of whether there is a different, more sustainable, approach to bee improvement. This book explores the alternative system of selecting and improving our stock from the bees that we already have. This approach immediately puts us into the position of being able to select bees for the qualities that we would like to see and for their performance under our conditions, rather than what does well in completely different circumstances.

The starting position

The first reaction to this approach is to assume that we have no stock worth breeding from. Immediately we are back to where we were before, of wanting to buy in stock of a higher standard. A difficulty of breeding bees is that we have so little control over the male line; queens will mate with 10 to 20 drones, or more, from over a six-mile radius. We are at the mercy of the quality of the drones in our area. However, any selection and propagation of queens that we do in the area will start to have an influence over the drone population. The more queens we raise, the more influence we can have over the drone population in an area, in the next generation. By working to improve our local population we can gradually increase our influence on the drones in the area and instead of adding to the problem, a start is made in bee improvement.

Many advocate bringing in some good stock to start with which will bring up the quality of the drones in the area. However the strain may be so different to the background population of bees in that area that it just adds to the problems of hybridisation, namely issues of temper and very variable stock. As the 'new' strain is inevitably lost, more bees have to be brought in as re-enforcements which gets us back to the same old problems.

This approach is not an 'instant fix' and, realistically, we will be starting from a low base in terms of quality. The temptation to import bees of seemingly good quality is great but ultimately contributes nothing to the quality of bees in this country, leads to disappointment and the vicious circle of endless imports.

Once we accept the starting position we can set off on the exciting adventure of bee improvement. It relies on a combination of natural and artificial selection. We work with 'natural selection' which selects stock for its ability to survive and into that we add selection by the beekeeper, 'artificial selection', for the qualities that we wish to see in our bees.

Natural Selection

Importation works against the benefits of natural selection. Nature favours individuals which can survive and reproduce. Weak or ill-adapted animals or plants will tend to disappear while those that thrive will be the ones which reproduce and allow the genes to continue into the next generation.

One of the great advantages of working to improve one's own bees rather than repeatedly bringing in new stock is that these are the ones which have survived under the prevailing conditions. Over generations they have had to prove that they are adapted to the environment that they are living in. By accepting the effects of natural selection the beekeeper is taking a simple but effective step towards producing a hardier bee.

The continuous bringing in of new genes that have not been tested in our environment negates the effects of natural selection and prevents us from reaping the benefits of this process. Through the avoidance of a steady input of new stock, weak or unsuitable colonies are 'weeded out' allowing us to work with, and select from, the bees that nature prefers. Long established colonies, with little or no influence from recently imported stock, may provide the best breeding material for individuals or groups interested in bee improvement.

The high death-rate of bees amongst beginners may be due to the use of bees which are unsuited to local conditions rather than the fault of the new beekeeper. The most readily available stocks of bees, for beginners, are nucs to which an imported queen has been added. Is it any wonder that losses are high?

Artificial Selection

Of course, the ability to survive, which nature selects for, although a very important attribute, is only one aspect, and beekeepers usually also look for other qualities in their bees, such as docility or honey production. Bees can be selected for whatever qualities we wish to see in them; if it is docility, we can rear queens from the docile colonies and eliminate the aggressive ones. This process can be thought of as 'artificial selection' and by using this in tandem with 'natural selection' beekeepers have a very effective tool for producing an improvement in the quality of their bees.

Since Neolithic times farmers have been selecting and improving crops and livestock. Can the same be done with bees or do they present special difficulties? One problem with 'breeding' bees is that the queen will often mate with 10 to 20 drones from a 6-mile radius, so full control of the male line is impossible for ordinary beekeepers.

The use of instrumental insemination (I.I.) has been developed and is used in professional bee breeding establishments. Some beekeepers regard this as a way forward but it is a hi-tech approach, being expensive on equipment and requiring high levels of skill. The aim of this booklet is to focus on a pragmatic approach and methods that can be achieved by the average beekeeper and so the use of instrumental insemination has not been considered here.

Bee Improvement and Bee Breeding

The terms 'bee improvement' and 'bee breeding' are often interchangeable but perhaps a distinction can be made. It may be more accurate to use the term 'bee breeding' when complete control over the male line (as well as the female line) is achieved either through the use of I.I. or by using a secure isolated mating apiary. 'Bee improvement' may be a more appropriate term for the position that most of us find ourselves in; we can achieve complete control over the female line (50% of genes) and aim to have an influence, though not complete control, over the remaining 50%, perhaps, for example, through the drone flooding of an area.

Although Improvement of our bees is made more difficult by our inability to have full control over the male line, we do, on the other hand, have the advantage that the honey bee allows the production of one generation per year so the rearing and selection process is a lot faster than for many other livestock.

Breeding true to type

Apart from selecting stock for the qualities that we want to see in our bees, the other aspect of importance is whether our stock breeds true; do the offspring reliably resemble their parents? What can we do to increase the chances of our stock breeding true and therefore making our progress more rapid? The aim is to get stability and consistency into the breeding population, just as breeding from pedigree dogs produces more consistency in the offspring than breeding from mongrel dogs.

Nature has responded to the differing conditions that the honey bee experiences over large geographical areas by the evolvement of numerous sub-species. Thus a central European bee (*Apis mellifera carnica*) has evolved to cope with the continental climate of very cold winters but reliably hot summers. The Italian bee (*Apis mellifera liguria*) is conditioned to the warm wet winters and hot dry summers of the Mediterranean climate. Our native sub-species (*Apia mellifera mellifera*), after millennia of natural selection, is at home in our rather fickle maritime climate of relatively mild, unsettled weather.

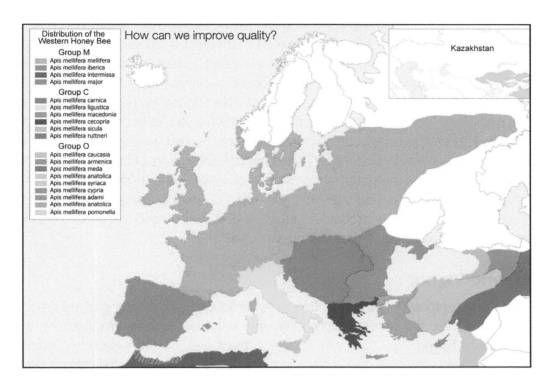

Figure 1. Distribution of sub-species (circa 1850) of the Western honey bee *Apis mellifera mellifera*
(Image: Karl Udo Gerth; creativecommons.org/licenses/by/3.0)

If we were able to turn the clocks back to the 1850s and look at the honey bee population in the British Isles, that is, in the UK and Ireland, we would see that all colonies belonged to the native sub-species *Apis mellifera mellifera* (See Fig. 1). After the last Ice Age, which ended about 10,000 years ago, bees would have colonised the area as the ice retreated and suitable forage became available. Within this population there would be considerable variation, or genetic diversity, as local bee populations adapted to their particular environment. Thus, for example, a honey bee from Southeast England would have different characteristics to one from Northwest Scotland or Southwest Ireland although they would all be of the same family or sub-species, namely the dark European honey bee. The genetic diversity, and therefore the variation in characteristics within a sub-species, allows us the opportunity to select for whatever qualities we would like to see in our bees.

This diversity within a sub-species is an important factor and means that instead of asking which sub-species has the best attributes for our needs, we should ask which sub-species would ultimately dominate in our area, without the influence of further imports, and therefore which would be the easiest one to select and maintain. It is easier to work with the forces of nature than against.

Despite 150 years of imports and the supposed demise of our native sub-species, due to the Isle of Wight disease, *A.m.mellifera* is still very influential in our bees. It is believed that an average colony of bees, unless of very recent import, still contains a high proportion of genes of the native sub-species. Exact figures for this will become available as the results of DNA analysis on honey bee samples are published. The relatively large influence of the native bee has been maintained despite little or no attention being given to it from the vast majority of beekeepers. Without the constant input of new genes through the import of other sub-species, and bearing in mind the apparent dominance of native drones in our conditions, it is reasonable to assume that, over time, the native bee would still naturally dominate in our environment. It is probable, though, that it would never revert to the pure strain of 150 years ago, except perhaps in a few areas.

Hybrid Vigour

Some believe that a more productive bee is produced by the mixing of the breeds as we then reap the benefit of hybrid vigour or heterosis. Indeed, in agriculture, the benefits of hybrid vigour are well known in plant and animal husbandry. Brother Adam was an advocate of this approach and felt that mixing or hybridising gave us the benefits of a more vigorous bee. In animal husbandry pedigree breeds are maintained which can then be deliberately crossed to produce hybrid offspring. This option does rely on the maintenance of pedigree stocks; it is a precise science, where distinct breeds or varieties are crossed and is much more difficult to replicate in the honey bee.

In bees the mating process is more complicated than in other livestock with multiple matings taking place with drones from a 10 kilometre radius. It is very difficult to produce a perfect cross between two distinct breeds, except through instrumental insemination or the use of reliable isolated mating apiaries which most of us do not have access to. This is practised on island mating stations off the coast of Denmark and Germany. Some breeders are producing queens which are a cross between two distinct lines of bees. It is a similar approach to Brother Adam's and they, therefore, call them Buckfast bees; the bees benefit from hybrid vigour and are intended for honey production; they are not designed to be used for selection and breeding as consistency in the offspring would not be produced. The crossing of F1 hybrids produces great variability, as shown by Mendel, and this failure to breed true is the reason gardeners do not save seed from F1 varieties. Of course, in the bee world we are not dealing with F1 generations but a hybridised mixture of bees which are constantly being added to by further imports. Thus, producing stock which breeds true is a big challenge for most of us.

The random nature of matings in bees makes the results of crosses unpredictable. Many of us have experienced the occasional bumper crop of honey from a colony which was probably produced as a result of hybrid vigour. If queens are reared from such a colony, the results are generally disappointing with no colonies matching up to the performance of the parent queen. Often the temper of the bee is also affected by hybrid vigour and such productive colonies can be impossible to work with, because of their aggressiveness. This is where the notion that 'aggressive colonies are the most productive' comes from. The main problem, though, is that this level of productiveness is just the result of a chance cross-mating and cannot be replicated in the offspring.

The use of hybrid bees may be an option for some who have the facilities to produce them, or for those who are happy to accept bees produced by others, usually with qualities designed to perform in quite different conditions to our own. Bee improvement should be on the agenda of all beekeepers and it should be possible, ultimately, to produce the best bee to suit the local conditions. If the principles of bee improvement are correct they should be applicable in every circumstance. Hybridised bees do not lend themselves to ongoing selection and improvement.

However, it may be possible to reap some of the benefits of hybrid vigour without the end result of totally uncontrolled mixing of different sub-species. This is something that breeders of any pedigree animal look for to keep a breed vigorous. Different strains within a breed, or sub-species, can be crossed from time to time to produce a similar effect to hybrid vigour. These strains, being of the same sub-species, would be unlikely to produce the aggressive colonies often associated with the hybridisation of certain sub-species. However, the maintenance of different strains, within a sub-species, may still present a problem.

Compatibility of sub-species

It is often suggested that a bee improvement programme can benefit from being able to select the best qualities of bees from around the world, as advocated by Brother Adam. He listed in detail the positive and negative effects on specific qualities, of crosses between different sub-species. Some sub-species are known to make apparently good crosses such as the Ligurian and the Carniolan, and are therefore considered 'compatible'.

Other crosses of bees can produce offspring with aggressive behaviour and are therefore 'incompatible'. Compatibility may be explained, to some extent, by understanding the evolution of the sub-species and how closely related they are to each other (See Fig.2). The species, *Apis mellifera*, the Western honey bee, is believed to have evolved in Africa and gradually colonised Europe and the near East. During this evolution, the honey bee has divided into geographic races and at least 28 subspecies have been recognised based on these geographic variations. According to work by Ruttner, and subsequently confirmed by analysis of mitochondrial DNA, each of these subspecies can be assigned to one of four major branches or lineages; African subspecies are assigned to branch A, North-west European subspecies to branch M, South-west European subspecies to branch C, and Near East subspecies to branch O (*See Fig.2*)

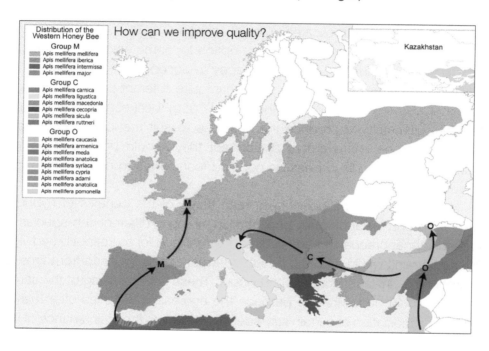

Fig. 2 Map showing 3 of the 4 lineages (European and Near Eastern) and the sub-species

Branch M: North Africa and West Europe (migrated via Gibralter)
Apis mellifera intermissa
Apis mellifera major
Apis mellifera iberica
Apis mellifera mellifera

Branch C: Central Europe and North Mediterranean coast
(migrated via Eastern Mediterranean and Turkey)
Apis mellifera carnica
Apis mellifera ligustica
Apis mellifera macedonica

Apis mellifera cecropia
Apis mellifera sicula
Apis mellifera ruttneri

Branch O: Near East and east of Black Sea
          (migrated via Eastern Mediterranean and east of Black Sea)
Apis mellifera caucasia
Apis mellifera armenica
Apis mellifera meda
Apis mellifera anatolica (western)
Apis mellifera syriaca
Apis mellifera cypria
Apis mellifera adami
Apis mellifera anatolica (eastern)
Apis mellifera pomonella

Sub-species within the same group are more closely related to each other than to those in another group, and appear to be more compatible. Thus the Ligurian or Italian bee (*A.m.liguria*) and the Carniolan (*A.m.carnica*) are closely related, both being in group C which developed from bees that migrated into Europe around the Eastern Mediterranean. Crosses between Ligurian and Carniolan appear to present few problems and are used to produce the hybrid Buckfast bee.

The Dark European honey bee *(A.m.mellifera)*, on the other hand, belongs to branch M, having colonised Western Europe by migrating around the western Mediterranean. This may explain the apparent incompatibility of crosses between *A.m.m.* (Group M) and either the Ligurian or the Carnican (Group C), which often result in bad-tempered or aggressive bees.

The compatibility factor is most clearly demonstrated in recently imported bees. If, for example, one buys in a Carniolan queen, the temper of the bees from these queens may be remarkably placid. The colonies produced will contain little or no genes of the native bee. When new queens are produced from this colony they will inevitably mate with drones from the local population and a proportion of genes from the native bee are introduced. The incompatibility, of these two sub-species is demonstrated by the rapid deterioration in temper that often occurs and may result in unmanageable colonies. Some beekeepers are quick to blame the native bee for this aggression but it is more accurate to see it as the result of hybridisation between incompatible sub-species.

However, if one perseveres with these bees for another generation or two, the worst of the aggressiveness disappears and the bees become manageable once more. It seems that compatibility between these aggressive colonies and the bees in the local population increases as the proportion of genes of the native bee, within these colonies, rises. One can, therefore, often work through the bad-tempered phase, as an improvement tends to return in subsequent generations. However, to get back to the ultra docility of the original imported stock, the beekeeper is often tempted to import again.

A similar process, though apparently more extreme, can be witnessed in the 'Africanised' or 'killer bee' of South, Central and now North America. When an African sub-species, *A.m.scutellata* (Group A), crossed with other sub-species of the Western honey bee in South America, originally from Europe (Group C), the resulting hybrid was extremely aggressive. This demonstrated the severe incompatibility between these types of bee. In this example, the temper also appears to gradually improve over subsequent generations.

What does this incompatibility tell us about our bees and about bee improvement? Importation may produce a short term gain in behaviour but generally has a negative effect on the temper of our bee population, from which it may take a long time to recover. This effect on temper is readily admitted by many advocates of importation, using it to explain why more importation is required in subsequent seasons. Others say that they have not noticed this deterioration and one can only assume that their imported queens happen to be compatible with most of the bees in their immediate area; perhaps theirs is an area with already high levels of imported bees.

The incompatibility, generally experienced when using imported sub-species, at least demonstrates that the genes of our native bee are still sufficiently present in the background bee population to produce this incompatibility. The level of influence of the native strain will vary from area to area but, whatever the level, unless non-existent, represents an important local resource, that through a selection and improvement programme, can be capitalised on.

Hybridisation and the development of a strain

The importation of queens of different breeds or sub-species, over the last 150 years, has left us with bees of very mixed genetic origins. This is true, not just in Britain and Ireland, but in many parts of the world. The end result of these ongoing imports is the generally low quality of bee that many of us experience and it is partly this that fuels the appetite for further imports in an attempt at making an improvement. Importation of queens may provide temporary relief for individual beekeepers but will not result in an overall improvement in our stock. The resulting crosses, of incompatible bees, not only leave us with bad-tempered bees but also with stock in which the queens do not breed true. It is difficult, therefore, for a selection and improvement programme to produce worthwhile results if imported queens are constantly being introduced into the system.

Rather than relying on hybrid vigour, to give us good results, with all the associated problems of incompatibility and unreliability in the offspring produced, the alternative route is to focus on selecting and improving from the stocks that we already have in our area. These stocks may be of very mediocre quality and we should recognise that we are starting from a low base. If we are working to the right principles, however, we will be able to see a steady rise in the quality of our bees.

To achieve reliability in our improvement programme we must return to the idea of selecting and improving within a strain and aim to improve the quality of the whole population of bees in an area. Bearing in mind our starting position, this presents a dilemma. Either we have to bring in bees of a 'pure' strain to start with or we aim to produce a strain from the bees that we already have. If we bring in bees of a pure strain, we are then faced with the issue of how to maintain that strain. On the other hand, if we aim to produce a strain from the bees that we already have, it may take a lot of time and effort, particularly if our starting stocks are very mixed. There is also a third option, which is a combination of the previous two, that is selecting from what we have got as well as adding to our stock by bringing in more pure-bred bees of the same strain. There has been much debate, within BIBBA, as to which is the correct approach and I think it is fair to say that all approaches have their merit.

The wisdom of using the native bee is not accepted by all, presumably, either because this type of bee is so under-represented in some areas or because they do not believe the qualities that we need in our bees can be found within the native population. All strains of bee are very diverse and we should be able to find and bring out the qualities that we want. Realistically, all beekeepers are not going to agree, but the more that are working towards the same goals, with the same type of bees, the easier it will be.

Buying in queens from overseas hinders a bee breeder's work.
Over 10 000 queens of non-native sub-species are imported into the UK every year. (J Phipps)

The aim is to reduce the variability within the population by narrowing down the genetic variation and thus creating a strain. Beowulf Cooper (1986) defined a strain as, 'a group of colonies which show uniform or fairly uniform characters and breed true to type'. He went on to say that the characters of parents and offspring, within a strain, are generally similar and that a strain can be maintained, as it is, or improved through selective breeding. Ruttner (1988) added to this theme by suggesting that by increasing the best stocks and replacing the bad queens with good ones, apiaries of uniform good performance could be created.

Selecting within a strain

If we accept that hybridisation of our bees is not conducive to bee improvement, the question arises of how we find the way forward from our current position. It is interesting to conjecture what would happen to our bee population if it was just subject to the forces of nature, that is, there were no more imports and no movements of bees within the country. (We will ignore the effects of Varroa which would obviously be a major factor, although we can assume that, in time, the bees will evolve to deal with this problem).

Many beekeepers notice the apparent dominance of native drones in our environment, with imported strains quickly reverting to more native appearance and characteristics. Irrespective of this, it seems likely that through the effects of natural selection, our bee population would gradually develop into or, perhaps we should say, revert to something resembling our original native bee. Presumably, given enough time, a relatively homogenous strain of bee would evolve which could probably best be described as 'near-native'. It may resemble our original honey bee but the chances of returning to the pre-import position of 1859 of a pure native strain may be remote. It may be, for this reason, that some in the bee improvement movement advocate the re-introduction of 'pure' native stock, believing that this will allow more reliability in an improvement programme. Others prefer to use stock which is proving itself in their area, believing that if the basic principles are adhered to, greater reliability, namely offspring breeding true, will naturally follow.

Working with the effects or tendencies of nature seems an obvious thing to do and offers us a greater chance of success in our quest for improvement. Selecting bees within a single sub-species seems the easiest way to develop a strain. BIBBA (The Bee Improvement and Bee Breeding Association) has long promoted this approach with the use of the native strain Apis mellifera mellifera.

The arguments for and against the use of the native bee can produce very polarised opinions, and nothing is gained in the process. Often one's views are formed according to the conditions that are found in a particular area; those with a good representation of near-natives tend to be more in favour of using native bees than those with little representation. Both sides are happy to shoot the other down but an effort to understand the other's position and perhaps reach a consensus may help us get over the current impasse where progress in bee improvement goes nowhere. It should not be us against them but rather how do we all move forward from whatever position we are in.

Risks of inbreeding

Inbreeding depression occurs when related animals breed with each other and the effect is the opposite of hybrid vigour. In honey bees the effects can be even more serious than in many other species, as the reduced number of sex alleles in an inbred population increases the chances of producing non-viable larvae. Homozygous larvae having two identical sex alleles produce diploid drones which are removed from the cells in the larval stage by the nurse bees. An uneven brood pattern, with many missing larvae, may be an indication of this problem.

By trying to select within a single strain, it is sometimes argued that the risk of inbreeding is increased. However, in a system which relies on the open mating of queens, even with the aim of flooding the area with particular drones, the mating process of the honey bee (i.e. multiple matings with drones from a wide area) makes the chance of the occurrence of inbreeding unlikely. Inbreeding may become a problem in a closed system, like a small island population or in a system which relies heavily on instrumental insemination, or when many daughter queens are derived from single 'breeder queens, but it is unlikely to present difficulties in an open mating system.

A certain amount of inbreeding can help to produce a more uniform population, which allows more consistency. In Dorian Pritchard's articles (Pritchard, D., 2011) on inbreeding he says, 'A community of bees should be sufficiently out-bred to be vigorous and disease free, but sufficiently in-bred to remain true to type'.

Use of the Native Bee

Despite endless imports over the last 150 years, and the reported demise of the native bee (from the 'Isle of Wight disease') colonies of native or near-native honey bees can still be found. The apparent dominance of the native sub-species, plus its obvious suitability to survival in our environment makes it a natural choice to select and improve. Although pure strains of native bee are unlikely to be found except in the most remote areas, it should be possible to re-establish a near-native strain through selective breeding.

The native bee has fared a lot better in some areas than others. This is largely down to the level of importation that an area has been subjected to over the years. Even areas covered by different Bee Keeping Associations can differ, according to what bee has been favoured by influential beekeepers in that area. It can be surprising, though, how examples of near-native bees occur in seemingly unlikely places, having survived, apparently against all the odds. There must be a mechanism which has allowed the native bee to remain, to some extent, as a separate sub-species and to avoid a complete loss of identity. This may be due to native drones being more likely to mate in our cooler, wetter, weather conditions. There may also be other factors to do with their mating habits which help to keep the breeds separate. The apparent dominance of native drones explains the difficulty of maintaining exotic strains of bee without the constant need to import new stock.

The survival of native or near-native bees, despite the level of disregard and lack of interest in the breed, means it may be possible to restore the native bee in any area of the United Kingdom and Ireland. It may represent a formidable challenge but by rearing queens and then selecting from those colonies most resembling the native bee, a more consistent strain can be achieved.

Some people may argue that by doing this we are committing ourselves to an inferior breed of bee, or that we are ignoring many good quality colonies that could be used in our breeding, and throwing away all the good breeding work that has been done in other countries. It may be argued that to restore the native bee is a step backwards and that we just need to select for the qualities that we want in our bees. I think this is a question that everyone needs to answer for themselves; is it possible to make progress in bee improvement and achieve consistent quality in our bees without moving away from a hybridised population and towards a single strain?

Diversity of Characteristics within the Native Strain

The argument in favour of using the native bee is not that it has superior characteristics to other breeds but that the process of bee improvement can most easily be achieved by using this strain. The main problem with the use of other strains is the continual deterioration of their qualities as they become hybridised, to a large extent by native or near-native bees. If we could stop this mixing of different sub-species, through reducing imports, we could then focus on the task of selecting and improving what we already have.

All of us have experienced that the qualities of bees vary enormously, and this is also true within each sub-species. For example, the native strain has evolved to be less prolific, on average, than the *Ligustica* as this is what has survived best in our conditions. However, within the native strain, there is enough variation to give us the option to select more prolific strains if that is of benefit to us. Similarly, the more docile and calm strains can be selected in order to produce a good tempered bee. Selection within the sub-species offers us the opportunity to adapt the bee to any of our requirements.

Working with 'The Best of What We Have Got'

Suppose we live in an area where there is little evidence of the presence of the native bee or it is overwhelmingly dominated by bees of another sub-species; where does that leave us; how do we proceed from such a position? Should we all be trying to 'import' pure native stock to start our bee improvement, or should we be working with and selecting from the current stock that we find in our locality? These two options, perhaps, represent two extremes of position and, I think, a consensus within the bee improvement movement has yet to be reached. Beekeepers have to assess their own circumstances and decide on what they feel is the best way forward for them.

I am inclined to say, 'Work with the best of what you have got', keeping a firm eye on, and assisting, any trends towards the native bee. My view is obviously coloured by the position that I found myself in, that is, much hybridised stock but the availability of some apparently 'near-native' colonies. Others may advise you to bring in some native or near-native stock to get you started. Many of us are not in the ideal starting position but bee improvement is relevant to all and not just for those in the most favourable circumstances.

Purity of Strain

For those attempting to use the native strain, how important is the 'purity' of the strain? The term 'near-native' is commonly used and it is recognition that most, if not all, of our 'native' stock will contain some genes of other sub-species. Our stock may look native and have the same characteristics as the native bee but at the DNA level it is unlikely to be the same as the pre-import native bee of the mid-nineteenth century. If we develop a strain that is adapted to our conditions, and that breeds true so that offspring resembles parents, that may be all that is necessary for bee improvement but perhaps not for those interested in the conservation of the native sub-species. Most of us are using stock in various states of purity, be it 10% or 90%. Our starting positions may be very different but through the selection process we can develop our strain over time and produce bees compatible with those of other beekeepers going through the same process.

It is easy to dismiss our own stock as worthless and, instead, look around for pure stock of the native strain to bring into the area. This may sound like a simple answer as a lot of the work to re-establish the native bee has already been done. However, being able to maintain a pure strain of bee in circumstances where other types of bee currently predominate will be a frustrating and probably impossible task without the use of secure isolated mating apiaries or the use of instrumental insemination. Maintenance of stock will come down to the use of regularly 'imported' stock (from other areas, if not other countries), a system we are trying to move away from. A better approach may be to accept our starting position, and through the selection and rearing of the best, build up a strain of bee which is sustainable in the area. Some argue that if the bees we are using are not 'pure stock' then progress in bee improvement is difficult or impossible. I regard it as a challenge to demonstrate that there is an alternative and more sustainable route to bee improvement than merely bringing in 'better' bees.

The remaining chapters of this book deal with the practical aspects of bee improvement, how to make a start and how to produce sustainable progress.

PART 2:

THE PRACTICAL ASPECTS OF BEE IMPROVEMENT

1.Making a Start: Assessment of Stock

The bee population in an area should be regarded as the initial resources. Bee populations are dynamic with colonies migrating through swarming and the movement of bees by beekeepers, as well as by drones migrating between colonies and perhaps travelling for several miles for mating. Those beekeepers who have brought little or no stock into their apiaries for a long period of time may offer the best chance of finding stock which could produce a sustainable local strain (see Chapter 3). This chapter, though, is primarily concerned with the characteristics that we would like to see in our bees, how to assess these qualities and, thus, how to select queens to produce the next generation.

If we have more than one colony, we are in a position to assess and select the most suitable queen for producing further stock. The greater the number of colonies, the more choice of stock we have to provide the next generation. By co-operating and working with other beekeepers in the area, we are able to increase the availability of resources.

We need to decide on the qualities that we wish to select for, and assess our colonies for those qualities. Environmental factors obviously exert a major influence on how well a colony performs but, in bee improvement we are aiming to improve performance through refining the genetic makeup of our stock, as well as just through good management. This is achieved through the selection of the best and the culling of the worst.

In practice, with regards to culling, resources can often be used elsewhere. Many of the queens which are unsuitable for use in a bee improvement programme, for whatever reason, may head perfectly good honey producing colonies. These colonies could be used in out apiaries away from the mating area, either for honey production or for providing bees for nuclei, as long as they are not imposing on someone else's improvement programme. Using such colonies for nucleus production is very beneficial, as unwanted queens are replaced by bees of the desired strain. Each season the least desirable colonies can be earmarked for this purpose, thus providing a useful resource and helping to raise the overall standards by providing homes for the offspring of our selected queens.

Environmental and hereditary factors

As beekeepers, it is up to us to provide the optimum conditions for our bees to perform well, such as a weatherproof hive, adequate food supplies and a satisfactory level of Varroa control. Our management techniques will vary, though, according to our personal preferences and priorities. If, for example, our main aim is to produce a bee tolerant of Varroa, we may be prepared to accept a higher Varroa population in our bees than otherwise, in order to assess our colonies' ability to cope. It may be that our preference is for a low management and a low input strategy. The bees which we select, as they do well under our particular system, may vary from those of another beekeeper who favours higher inputs of time and resources.

It may be difficult to distinguish between the influence of environmental factors and hereditary ones. In the spring, for example, some colonies expand rapidly while others build-up more slowly. With a little help from the beekeeper in the form of stimulative feeding, seemingly poor colonies can sometimes leap forward. Are they weak because of an environmental factor (shortage of stores, for example) or are they genetically inclined to this slow start? A slow start may result in no spring crop, but if it means no swarming and a good summer crop it may be an advantage. On the other hand most of us need our bees in a condition to take advantage of nectar flows whenever they may occur.

If a colony produces wall to wall brood in August, we may be impressed by the queen's prolificacy (and perhaps the potential for the nectar flow from the heather) but, for many of us, the result is a colony with an unnecessarily large population of bees for the winter, which merely means the consumption of more stores. We should try to judge bees on the end results rather than on what is impressive to the eye.

In order to work out which of our colonies have the most desirable traits, and therefore which of our queens have the best genes from a breeding point of view, we need an accurate system of assessment. By not repeatedly bringing in new stock and by using bees which over-winter well, we are allowing natural selection to play a big part in the choice of our bees. Hardiness is already being built in as a quality as the non-hardy bees die out and so play no further part in the selection process. We now need to refine the process further by assessing our bees for the other qualities that we would like in our bees.

Selecting characteristics to assess

To assess the quality of our stocks for the characteristics that we are looking for we need a marking or scoring system which allows us to simply and accurately record their qualities. The design of an assessment system should be guided by the following criteria:

- Selection of fewer qualities will offer a greater chance of producing an improvement in those qualities
- Assessment should require minimal time and effort
- The marking system should be simple, for ease of use and objectivity
- Qualities selected should be relevant to the characteristics that we wish to see in our bees, i.e. we have to decide which qualities are the important ones.

The five qualities that BIPCo assess are:

- Native Appearance - (to help us work within a strain)
- Temper
- Low swarming
- Health and brood pattern
- Productivity

Ultimately one could just select for good temper and productivity. 'Low swarming' and 'health' are merely factors that will impact on productivity. However, we feel that these qualities require special attention as they are so important to productivity and ease of management, and the development of a quality bee.

Temper

This quality is high on most beekeepers list, particularly beginners. Even for experienced beekeepers good temper has an effect on the ease of management, the pleasure of beekeeping and the safety of the general public. For these reasons we consider good temper of the utmost importance.

Whilst the temper of the bee is ultimately governed by genetics, environmental factors can have a major influence. For example, temper can vary according to how bees are handled, the weather, available forage, the time of year and by different stages in the swarming cycle. An assessment is made every time the bees are handled and, over time, an average score for the colony will result. If the bees are bad-tempered on occasions, but are normally fine, that will show up in the records, and will not be regarded as a problem. Similarly colonies that are persistently aggressive would be highlighted in the records and that would be seen as an issue that needed to be addressed.

Temper is marked on a 1-5 scale, with 5 being the best score.

5   bees exceptionally docile and noticeably calm with little flying or movement on comb

4   colony easy to handle with no aggression but some running on comb, more flying and more excitable but no aggression

3   very nervous (hanging in clusters on comb), rapid running, lots of flying and/or some aggression and stinging

2   aggressive, lots of stinging, but little or no following when the hive is closed down

1   very aggressive, stinging and following. Generally unworkable.

In practice, a mark of '4' is given to an average good-tempered colony, and '5' is reserved for colonies with exceptionally docile behaviour.

Low swarming

Low swarming is perhaps the single most important trait in improving honey bee productivity. Swarm control methods are very time consuming and, even if employed successfully, result in some loss of productivity. All too often things go wrong and a greater loss of productivity is experienced. For maximum honey production and ease of management, the swarming impulse needs to be at an absolute minimum.

Bees appear to have become increasingly swarmy since the advent of Varroa. At one time it was thought to be the norm that an average of one colony in three would swarm each season whereas, these days, many beekeepers experience a swarming rate close to 100%. Swarming allows a break in the brood cycle and therefore is a useful mechanism for reducing the reproduction of the Varroa mite. Colonies which swarm will support a lower mite population than those with continuous brood and therefore will be more likely to survive. Non-swarming colonies allow uninterrupted growth in the mite population and are more likely to succumb to the effects of Varroa. Natural selection has, therefore, favoured bees which swarm.

In the light of the Varroa problem, swarming is an example of where the beekeeper's interests may be against those of natural selection. Any reduction in the propensity to swarm will have to be accompanied by efficient Varroa control by the beekeeper or an increase in the natural ability of the bee to cope with Varroa.

The swarming process is, of course, very much a natural part of the honey bee's make up as it is its means of reproduction and, thus, crucial to the survival of the species. Swarming allows an increase in colony numbers which makes up for

any losses which may have occured. It is in the interest of a pro
swarm in order for the genes to go through to the next gener
traits there are many environmental factors which affect the swar.
However, it is a fact that the bees' propensity to swarm varies enormously be.
colonies; some, for example, may prepare to swarm after having just filled the
brood box with bees, whilst others may fill a couple of supers of honey before
swarming. It is this genetic variation which allows selection and improvement by
the beekeeper.

The production and distribution of queen substance within the colony plays
an important role in the swarming impulse. Young queens produce more of this
than old queens but, presumably, the length of time that enough queen substance
is produced to suppress swarming varies from queen to queen. A colony which
goes through the first full season of the queen's life without attempting to swarm
is beneficial to the beekeeper, both, for ease of management and for honey
production. If we use 'bought in' queens, we have no hope of increasing our
chances of achieving this. By selecting our own queens we can rear from non-
swarming colonies and thus encourage a genetic propensity for less swarming.

Assessment of a colony for low swarming is, perhaps, less straightforward
than assessment for temper, but again the aim is to register results on a 1-5 scale.
It may be a good idea to combine the marking with an assessment of 'colony
strength' in comparison with other colonies in the apiary. It would be necessary
to note the size of each colony on inspection and after all had been assessed a
mark of 1-5 could be awarded according to the average for the apiary; an average
strength colony would receive a mark of 3, and others marked in comparison with
this. This would provide useful information as to how the colony over-wintered
and how it is building up through the season. A colony which swarms would be
reflected in the marks by a sudden loss of colony strength.

Regardless of colony strength, the fact that a colony has swarmed would count
against it and it would be downgraded to a 1 or a 2. If, on inspection, a colony is
found to have made preparations to swarm (i.e. is producing swarm cells), although
not actually swarmed, this colony would also be marked down, so it may go from
receiving a '5' for colony strength to, for example, a '2' for attempting to swarm.
The important score is the last one recorded, rather than an average of results.

The scoring for 'low swarming' attempts to make allowance for mitigating
circumstance', hence:
- A colony which makes no attempt to swarm during the season, and is
  productive, could be marked as 5.
- A colony which makes no attempt to swarm but is less productive could be
  marked as 4

- A colony which only attempts to swarm after it has been productive (say, filled 2 supers) could be given 3
- A colony which swarms after building up to strength but without being productive could be marked as 2
- A colony which attempts to swarm before apparently building up to full strength (i.e. swarms as a small colony) could be marked as 1

Supersedure is regarded as a good trait. Productive colonies which change their queens in this way are highly regarded and could score 5.

NB:
Supersedure may, however, be more prevalent during the latter part of the season in colonies headed by an old queen who has, in fact, swarmed earlier in the season
- supersedure does not offer a perfect strategy for the bee in the wild as no colony increase occurs and so there are no replacements for any natural losses

Brood pattern/health

For simplicity these two characteristics have been linked under one heading although, perhaps, they may have little in common. In fact, one could ask why brood pattern is regarded as important at all. Does it have any impact on honey production, for example, or is it merely cosmetic? A good brood pattern gives the impression of being a good queen, and while this may not be totally true, good impressions can be important. A poor brood pattern could be due to:
- A poor laying queen
- Chalk brood (or a more serious disease problem)
- Inbreeding (which produces non-viable larvae)

A good brood pattern is, therefore, a sign of health, the absence of inbreeding and a sign of the potential of the queen to build up a strong colony.

solid blocks of brood are preferred but it is also desirable to see combs with an arch of pollen and honey around the brood nest. It is felt that this represents a more sustainable bee than one in which most of the combs are filled with brood and no stores. In times of dearth, colonies without the 'cushion' of stores will be the ones most prone to starvation. Of course this trait can be affected by the size of the brood box used.

As far as assessment of the health of the colony is concerned, those with high levels of chalk brood or the presence of sac brood or bald brood will be marked down. Similarly, a colony with a higher Varroa count than others in the same apiary, having had the same treatment, could be marked down. The symptoms of the adult bee diseases of acarine and nosema would also reduce the marking. There appears to be a strong genetic link to the ability of some colonies to withstand diseases which others may be prone to.

Brood pattern/health is also assessed on a 1 to 5 basis, with an average score being 3, and 5 being the top mark. One would generally mark a colony as 3 unless evidence of something better or worse than average is noted.

Productivity

Ultimately we are aiming for bees which are not just good survivors but ones which can produce a good crop of honey. A colony which produces an enormous crop of honey in the occasional season when conditions are perfect may not necessarily be the best average producer. A better average production over a series of, more typical, mediocre summers may be achieved by another colony. Thus keeping bees which produce a terrific yield in a rare good season but which need constant feeding in poor to average seasons is probably not the best solution. Some colonies seem able to produce good crops of honey even in very unfavourable seasons and these may be the ones we should be encouraging.

By comparing colonies with the apiary average, a figure can be arrived at which allows a comparison to be made with colonies in other apiaries. Thus the ratio of each colony's yield, compared to the average yield in that apiary, can be calculated. This is achieved by:

- Estimating or weighing honey as removed, and adding up the total yield for each colony for the season
- At the end of the season, work out the average honey production per colony for each apiary
- The total yield of each colony can be compared with the average yield for that apiary
- For example, if 6 colonies produce 300 lb., apiary average is 50 lb.
- If a colony in that apiary produces 100 lb., it will get a score of 2.0, (i.e. 100÷50) i.e. twice the apiary average
- A colony producing 40 lb. will score 0.8 (40÷50)

Thus:

Colony rating =
Individual Colony Production ÷ Apiary Average Production

This colony rating is more relative to colonies in other apiaries than a straightforward comparison of actual yields. A colony producing twice the apiary average would be considered better than a colony in another apiary producing the average apiary yield, irrespective of the actual amounts of honey produced by each colony.

2. Assessment of Stock, Part 2: Selecting within a strain

'A community of bees should be sufficiently out-bred to be vigorous and disease free, but sufficiently in-bred to remain true to type'.
Willie Robson, Chain Bridge Honey Farm, Northumberland

The Problem of hybridization

Assuming the principles of bee improvement are true, they should work for beekeepers in any situation. One of these principles is that bee improvement can most readily be achieved when selecting within a single strain or sub-species because, as explained in Chapter 1, these bees are more likely to breed true from one generation to the next.

Bee breeders around the world generally confine their efforts to working within a single sub-species, just as other stock breeders keep flocks or herds of pedigree animals. Selection within a single strain, and sometimes even within a closed population, as in the New World Carniolan Programme, is generally the norm. Maintenance of genetic diversity within the strain is essential; even more attention has to be paid to this if working within a closed population.

Objections to this approach come from those who fear the loss of genetic diversity that, they believe, will occur through limiting selection to a single sub-species. Also many believe that the crossing of different sub-species is essential to achieving a better performance from their bees. The greatest advocate of this method was Brother Adam who stressed the importance of hybrid vigour. The 'Buckfast Bee' as devised by Brother Adam was a cross or hybrid of different sub-species. There are still advocates of this method, namely the crossing of two sub-species, and these breeders call the resulting bee a 'Buckfast'; it is usually a cross between Carniolan and Ligurian, which being closely related sub-species and compatible, produce prolific but gentle bees.

The 'Buckfast Bee', being a hybrid, is not a good starting point for the selection and improvement process as the diverse genetics of a hybrid bee mean that the offspring will be variable and unreliable. Brother Adam advocated that specific qualities could be selected from different sub-species and then, through careful breeding, be 'fixed' in the resulting bee. This fixing of the best qualities of different strains of bee is a good approach but something that can most reliably be achieved by working with different strains within the same sub-species.

Interestingly, Brother Adam did not condone the use of bees of unknown mixed parentage, generally referred to as 'mongrels'. In fact he went as far as to say that the 'greatest danger facing nearly every race of honey bee is the indiscriminate use of mongrel stock'. He pointed out that, to reap the benefits of hybrid vigour, crosses should be made between pure strains or sub-species of bee. What seems to have been overlooked is that the hybridisation of bees, that is, the crossing of bees of pure strains, that Brother Adam favoured, ultimately leads to the mongrelisation, or random mixing of our stock. Hybrids do not breed true and the policy of hybridisation puts the average beekeeper in a weak position with regards to bee improvement. The only option then becomes, as advocated by Brother Adam, the buying in of new stock. Thus, we return to the old cycle whereby real progress in stock improvement becomes impossible. It is beyond the bounds of most beekeepers to be able to maintain distinct sub-species and be able to cross them at will.

Development of a 'pure strain'

If we accept that the establishment and maintenance of pure stock offers the best chance of making progress in bee improvement, we need to look at how to develop or maintain a strain? Oddly, it is at this point that 'bee improvement' becomes controversial, perhaps because beekeepers do not like to be told what breed of bee they should be using. The problem is that if we all choose different sub-species the end result will be more mongrelisation.

Controversy, in the choice and establishment of a strain, may occur because we are all approaching bee improvement from a slightly different starting point and therefore resent the suggestion that one approach fits all. If we start by considering which type of bee would ultimately dominate in our area without the constant input of imports, most of us would conclude that it would be a native or, more likely, a near-native bee. It is unlikely that all genes from non-native sub-species would ever completely disappear from the bee population, even through a rigorous selection programme. It seems that native/near-native drones tend to dominate, probably largely because of cool-weather flying, and there may be other reasons to do with the mating habits of different sub-species. Instead of working against the natural inclination of the honey bee, it makes sense to work with it, and to use it to our advantage to raise the quality of our bees.

Early results from DNA analysis carried out on random samples of bees from around the country indicate that the dominant sub-species is indeed the native one, *Apis mellifera mellifera*. The easiest way forward seems to be to select and improve within this strain. To go down the route of opting for a different sub-species will be going against the natural tendencies and will merely increase our difficulties.

Numerous other sub-species are represented, in smaller proportions, in our bee population and although it is unlikely that we would ever be able to get back to a completely pure strain of native bee the aim must be to get back to a bee that breeds true. Whilst genetic diversity within the strain is imperative, both to avoid inbreeding and to give us the choice of all the qualities that we wish to select for, our problem is more likely to be one of too many genes from other strains entering the system rather than a lack of genetic diversity. This introgression, or infiltration of the genes of one sub-species into the gene pool of another, makes maintenance of a pure strain difficult but not impossible. The more beekeepers that work according the principles of 'natural' and 'artificial' selection, and refrain from importing exotic sub-species, the greater are the chances of raising the quality of our bees.

Applying 'bee improvement' principles in any circumstance

Beekeepers in other parts of the world will need to reach their own conclusions as to the establishment of a strain in their area. If, for example, you are a beekeeper from North America, where *Apis mellifera* is not native, there may be no obvious sub-species to choose. Even a beekeeper in some parts of England, may have little or no evidence of the native sub-species in their area, so may wonder where to start. The answer must be to select and improve from local stock and go with what tends to dominate in the locality, without the constant import of bees.

Whilst the principles of bee improvement should apply to beekeepers in all circumstances, difficulties arise if other beekeepers in an area do not co-operate and continue to bring in new stock. This is, to some extent, always going to happen. It increases the challenges of bee improvement and a lot of perseverance may be required as steps backwards can seem to be almost equal to steps forward.

For most of us, in attempting to establish a strain, the difficulty lies in the fact that our bees are often the end-product of this random mixing of different types of bee. Queens reared from our best stocks will inevitably produce very variable offspring and this is compounded by the mating of our new queens with the mixed drones in the area. Far from experiencing the advantages of hybrid vigour, we end up with poor quality bees where, it may seem, that the only answer is to bring in something better, thus perpetuating the cycle of the importation of more queens.

Starting from 'what you have got' versus 'bringing in pure stock'

Beekeepers have to assess the situation in their own areas and work out the best way forward. Many advocate the bringing in of pure native stock from which a native population can be re-established. This may provide an opportunity to start with some purer breeding stock and so save the time and effort which will be needed to develop a good local strain. However, new stock may not be locally adapted and the position may not be very different from those who like to bring in queens of other sub-species; the quality of stock may, similarly, decline over time, as introgression occurs, and further 'imports' (from another part of the country) become necessary.

It may be just as valid, and perhaps stand more chance of success, to resist the temptation of bringing in 'pure-stock' and work with the bees that are dominant in the area, bearing in mind that the largest genetic component of those bees may well be *Apis mellifera mellifera*. Without the constant influx of new genes and through careful assessment and selection, the population will have a chance to settle and tend towards homogenisation. Over time, it is likely that the near-native bee will dominate and from this a better standard of bee can be selected. The use of long established stocks in an area, will allow the selection along the lines of what is locally dominant, and a local strain can be developed.

As Brother Adam pointed out, 'because of multiple mating ..... we have queens which are at one and the same time pure mated and mixed mated. The progeny of such mixed mated queens is partly pure and partly crossed. Hence by careful selection we can breed queens of absolutely pure type from differently coloured descendants. This conclusion is derived from theory but confirmed by actual results' (Brother Adam 1987). Thus even if we have to use a hybridised breeder queen (as we have nothing else), a certain percentage of the offspring will be of the type, or close to the type, that we are looking for. By repeated selection over generations we can reduce the hybridisation and move closer to a near-native bee.

Identifying the Native Strain

If we are to select within the native strain we need to be able to recognise a native or near-native bee. Beowulf Cooper, (1986), has suggested what he considers the typical behavioural characteristics of the native bee, and other sub-species can be defined in a similar manner. We not only wish to select bees because their behavioural traits are typically 'native' but we also wish to select characteristics that are beneficial to the beekeeper. The huge range of characteristics within a sub-species is of interest to us from a bee improvement point of view; we are then able to pick and choose the qualities that we require. We do not need to rule out colonies from our breeding programme because certain qualities may not be typical of the breed. It is a matter of selecting our bees for the required qualities whilst at the same time recognising that our bees are of a particular strain or sub-species, namely, in our case, *Apis mellifera mellifera*.

There are three ways of recognising a sub-species, irrespective of their behavioural traits; by appearance, through morphometry, and by DNA analysis. This latter technique may provide us with the most accurate information but is generally beyond the means of the average beekeeper being technical and expensive. Its use is becoming more widespread, and it may become cheaper and more available in time. DNA analysis is allowing us to learn more about the origins of our bees and the way bees from different areas are related to each other. Samples of bees of native appearance have been taken from sites throughout Britain and Ireland for BIBBA's 'Project Discovery'. It is hoped that the results of their DNA Analysis will help us to understand more about the honey bee's 'family tree' throughout the British Isles as well as the relative purity of the sub-species that remains.

The use of morphometry

The use of morphometry is a bit less refined than DNA analysis but can give us a more scientific approach to recognition of the different sub-species than merely judging the bees' appearance. Historical samples of bees, from the time before any importations into this country, have been morphometrically tested and these have confirmed the apparent accuracy of the methods, although the accuracy of results from the testing of modern day hybrids is still being debated and may become clearer as more DNA analysis is carried out.

Further work comparing assessments of colonies by appearance, morphometry and DNA analysis will increase our knowledge of the usefulness of each technique. If all three methods are accurate, good correlation between the results will be seen. In my experience a colony with bees of a uniform native appearance correlates well with morphometry results for that colony, suggesting that the bees are, in fact, of the native sub-species or, more accurately, near-native, as many other sub-species (in smaller proportions) are likely to be represented in the genes. Early results of DNA analysis seem to confirm that, in colonies of bees with native appearance, the largest proportion of the genetic makeup are indeed of native origin but genes of other sub-species are present in lesser proportions.

Morphometry is a technique used in biological sciences to distinguish between similar species or sub-species. In the case of the honey bee, many different measurements have been compared in order to reliably identify the main sub-species. It is generally accepted that two most simple and effective measures are those taken from the veins in the right forewing of a sample of bees. These measures are known as the 'cubital index' and the 'discoidal shift' and by plotting one against the other on a graph, for each bee in the sample, a picture emerges as to what type of bee the sample comes from. The results can reveal which sub-species the bees are descended from and how much hybridisation has occurred.

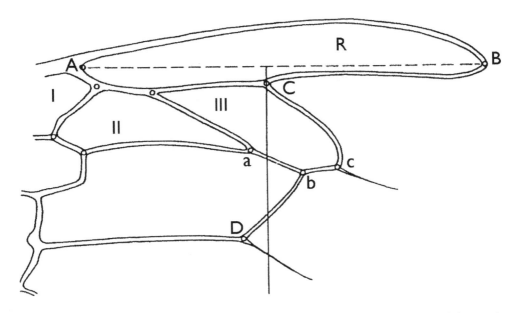

Fig. 3 Right forewing of worker bee illustrating 'cubital index c.i.' and 'discoidal shift DS (see text below)

Cubital Index

The 'cubital index' (c.i.) is the ratio of the length of the vein ab in the right forewing compared to length of bc.
    Thus c.i. = ab ÷ bc

Discoidal Shift

To calculate the discoidal shift (DS) a horizontal line is taken across the full length of the radial cell (R), that is from A to B.. From this line another line is taken at right angles i.e. vertically to pass through joint C. The relative position of joint D to this vertical line is the critical measure. If D is positioned to left of vertical line, it is considered a negative discoidal shift (a characteristic of *A.m.mellifera)* and if to the right it is a positive discoidal shift, characteristic of *A.m.carnica* and *A.m.ligustica*. The degree of positivity or negativity can be measured or calculated automatically when using appropriate computer software.

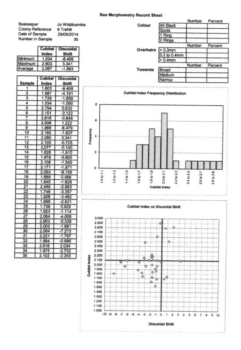

Fig 4 Examples of results from near native bees

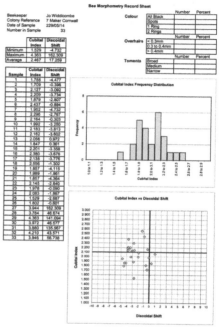

Fig 5. examples of results produced by different sub-species. A hybridised population produces dots scattered over a wider area on the graph whereas a tight cluster of dots represents a uniform population of bees ideally suited to producing uniform offspring.

'Interpretation of Morphometry Results'

A colony of bees of a uniform strain or of a particular sub-species would be expected to produce a cluster of dots in one part of the graph produced by plotting $c.i$ against DS for wing results. On the other hand a hybridised colony with two or more subspecies in its ancestry will produce a diverse scatter of dots on the graph. The principle is that breeding within a strain reliably produces offspring similar to the parents and therefore more readily lends itself to selection and improvement.

The ancestry of the colony being analysed is illustrated by the position that the dots fall on the graph. For example *A.m.mellifera* has a negative discoidal shift and therefore will fall on the left hand side of the graph. *A.m. Ligustica* and *A.m.carnica* have positive discoidal shift and so fall on the right side of the graph.

A.m.mellifera has the lowest cubital index in the range 1.5-2.1 so falls lowest down on the graph. *A.m.ligustica* at 2.0-2.7 will be higher (and on the right) whilst *A.m.carnica* (also on the right) is higher still at 2.4-3.0.

The results indicated in Fig. 4 (over) show a wide scatter of dots and illustrate a hybridised colony. This colony would not generally be considered for use in a breeding programme although it is possible that its performance as a honey producer could still be very good. The results illustrated in Fig.5 show a more uniform colony which could be useful in a breeding programme. Morphometry has no bearing on the performance of a colony but is a useful tool in selecting stocks which will breed true. It should therefore be used in conjunction with assessments of the performance of colonies.

The morphometry technique

A sample of bees needs to be collected from the colony to be tested. An average sample would be 30 bees although 15 may give a guide. Some users prefer samples as large as 60, or even 75, as this is thought to provide a more accurate picture of the genetic make-up of the colony.

Drifting of bees between colonies can produce inaccuracies in the results as the sample may contain bees from neighbouring colonies. To minimise this, samples of bees are best collected from a brood comb (avoiding the queen!) as these will contain a higher proportion of young nurse bees and therefore likely to include fewer drifted bees.

Samples are taken in the same way as collecting bees for disease diagnosis. The usual method is to slide an open matchbox over the bees on the comb to scoop the bees up and then slide the cover over them. The sample should be frozen for 24 hours and then rinsed in water and allowed to dry out. The right forewing of each bee is taken and laid on the bed of a scanner, or on a glass slide, taking care that no wings touch each other. An image is taken and saved on to the computer. A clear image is required as blemishes or marks on the wing can interfere with the results, hence the need to rinse and dry the samples.

The image of the wings produced is then analysed using one of several available computer software programmes. These programmes calculate the cubital index and discoidal shift of each wing. 'DrawWing' or 'Beemorph' are commonly used to produce results that are then fed into the software programme, 'MorphPlot'. This converts the results into graphs for easy interpretation (as in Figs 4 and 5). These programmes are available for download on the internet   . (see appendix).

N.B. Some operators prefer to use Beemorph as the wings can be attached to the slide or scanner using tape, such as masking tape, whereas for DrawWing the wings must be unattached and separate.

Identification of native honey bee by appearance

Even within a sub-species the appearance can vary considerably but one can get a definite impression of the type of bee by scanning one's eyes across a large number of bees on a frame. This may be the least accurate and perhaps the most subjective of the methods but once one gets used to assessing bees in this way it seems to provide a quick and fairly accurate guide.

The appearance of the native bee is hinted at by some of its common names, such as the 'dark European' honey bee and the 'black bee'. It is normally characterized by a dark brown or black abdomen and no yellow/orange bands at the top or anterior of the abdomen, which is typical of bees of Italian origin. The body hairs around the thorax are yellow/brown (ginger), as compared to the grey of Carniolans, and because of this people sometimes get the impression that the bees, when looked at en masse, particularly in bright sunshine, look yellow rather than black.

As we look at the bees on the comb we need to scan our eyes over the abdomens, particularly the anterior, that is, the part nearest the thorax, to assess the colour. One, two or more yellow bands or, more accurately, orange bands, indicates that the original source of bees may have been Italian (Ligustica). These could be from the Mediterranean or from the Buckfast strain, New Zealand or Hawaiian, as all have a strong Italian influence. Most colonies may be a mixture of dark bees and orange banded bees indicating the hybridised nature of many of our bees. When selecting for the native strain we look for uniformity in the bees, so we like to see colonies with few or no bees with orange bands, in the belief that the influence of exotic strains in these colonies will be minimal.

Unfortunately, distinguishing between the native bee and the Carniolan (A.m.carnica) by appearance is slightly more difficult. This bee may also be dark, but generally greyer and browner, than the native bee. The Carniolan is gaining in popularity, not just because imports are cheap and readily available but because it produces good tempered colonies with a very prolific queen. This may sound ideal, to some, but they have a reputation for being extremely 'swarmy' and after a generation or two of crossing with local drones, frequently become very aggressive and difficult to manage.

The Carniolan has a more slender body shape than the native bee and, although dark, is distinguishable in other ways. The overall impression is of a grey bee; the hair around the thorax is likely to be grey, and not yellow/brown, and abdomen colouring can be grey/brown near the thorax. The tomenta, that is, the light coloured bands on the 3rd, 4th and 5th tergites, towards the tail of the bee are grey and wide. In the native bee these tomenta can be grey but are narrow.

In the Carniolan, the tomenta may be 50% of each tergite, or segment, giving the impression of an even pattern of light and dark stripes towards the tail.

Hybrids between the Carniolan and the native bee become more difficult to distinguish but behavioural characteristics may help to spot these. A Carniolan hybrid, that is when crossed with the native bee, may be excessively prolific, swarmy and aggressive.

Fig.6a. *A.m.mellifera* queen and workers

6b *A.m. Ligustica* queen and workers (J Phipps)

6c Buckfast queen and workers (J Phipps)

6d *Apis mellifera carnica* (J Phipps)

Selecting for appearance or for morphometry

The judgement of bees by their appearance, or the use of morphometry are merely tools to help us to develop or maintain a strain. In themselves, they are meaningless as the appearance, or the measurement of veins in the wings have no bearing on the bees' performance. It is theoretically possible to produce a colony of bees that have the appearance or the morphometry of native bees, but are genetically of another sub-species, or perhaps more importantly, have the 'right' morphometry or physical appearance but have no other redeeming qualities as far as the beekeeper is concerned.

It is vitally important, therefore that we do not get caught up in attempting to create the perfect looking bee or the bee with perfect morphometric measurements to the exclusion of all other qualities. Morphometry and appearance should only be used in conjunction with the qualities that we wish to select for. On the other hand, both of these methods, if used sensibly, can be of enormous help in establishing or maintaining a strain. If we are able to develop a uniform appearance in our bees, or in our queens, we will be able to spot, at a glance, when a rogue mating has occurred with drones of another sub-species.

This method is not without precedent; Brother Adam was known to pick out queens that did not look right for the crosses he was making. Also, in the 1960s, a national bee breeding project was carried out in Egypt in which the Carniolan bee was introduced. This breed was partly chosen because its body colourations differed from the local bee and so it provided a genetic marker which showed up in the progeny if any of the new queens mismated with local drones.

Assessment for Record Keeping

Colonies can be assessed for native appearance on a 1 to 5 scale by looking at the bees on the comb and estimating what percentage look like a native bee. Yellow bands indicating a Ligurian influence or any indication of Carniolan such as the different body shape and the impression of grey colouring or light and dark bands or stripes in equal proportions on the abdomen would be marked down.

The characteristics of the colony may also give an indication of native or non-native, such as the extremely prolific nature of some colonies. Solid blocks of wall to wall brood may point towards a Carniolan or Ligurian influence.

Thus the marking system is:

5   Colony consists of bees of uniform native appearance, i.e. dark with little or no bees with orange bands on abdomen, or evidence of Carniolan influence (i.e. wide tomenta, or light coloured bands on abdomen; white/grey body hairs on

thorax instead of ginger or yellow/brown)

4   Fairly uniform with small minority of bees with orange bands or Carniolan (>10%)

3   A larger proportion of 'yellow-banded bees' or Carniolan (10-25%)

2   Very mixed population of 'Italian', 'Carniolan' or native

1   Fairly uniform 'Italian' or 'Carniolan'. No evidence of native influence

This is intended to provide a rough guide to be adapted to one's own personal circumstances.

Summary

DNA analysis of bees should be able to provide us with the most accurate assessment of which sub-species a colony belongs to as well as the extent of any hybridisation. This technique is not readily available at present but may become more widely available in the future.

Morphometry techniques can provide us with a useful guide to the purity of our stock. The technique should be used in conjunction with assessment of colonies for their qualities and not as a stand-alone selection technique. Further work comparing results of DNA analysis with morphometry results will reveal just how accurate and useful morphometry techniques are.

Assessment by appearance is the only method readily available in the field and can give us a useful ready guide. Its limitations are that it does not tell us the amount of hybridisation within our bees. Although they may look native, this may be hiding the fact that genes of other sub-species are present, particularly if the genes for native appearance are dominant. A combination of assessment by appearance and the use of morphometry can help us to fine-tune and recognise when hybridisation is occurring.

The appearance of our stock is not just a cosmetic exercise; it can offer us the chance to gauge when hybridisation has occurred. If our stock is of uniform native appearance, we would notice any changes which occur after requeening, for example, the appearance of yellow-banded bees would indicate that hybridisation has occurred. Assessment by appearance can, therefore, be a useful tool in the efforts to maintain a strain. It is also worth assessing the appearance of the queens and perhaps selecting for a particular look. Although appearance is irrelevant to the characteristics of the bee, it may help us to maintain a pure strain by allowing us to easily identify hybridisation; it, thus, takes on a new importance.

3. Record keeping

In the previous two chapters we have looked at the assessment of colonies for their performance, and also at the use of their 'appearance' as a tool to help to breed within the strain. These assessments are an on-going process during the active season when a lot of data can be collected about each colony. As the data builds up, it tells us a lot about the qualities of our queens and the suitability of each queen for breeding. If we are operating alone with just a few colonies it is possible to get an instinctive feeling for which queen to rear new stock from. However, as the number of colonies gets larger and particularly if a number of beekeepers are working as a group on bee improvement, accurate records are essential. A good system of record keeping will allow us to make an objective assessment of the qualities of each queen and thus select breeding stock on that basis.

Good records kept over a number of years may also help to reveal trends and indicate what progress is being made in bee improvement. Is the average temper of colonies, for example, better now than it was five years ago? Computerised records have the advantage of being able to digest large quantities of data and can help to reveal trends that may be impossible to spot in 'pen and paper' records. Also computer programmes can automatically carry out calculations when compiling the weekly records, producing averages, etc.

Whether using computerised or manual records, the initial recording is usually done manually, perhaps on a 'field record sheet'. This allows assessments for every colony to be jotted down after each inspection in the apiary. On return, the data can then be transferred to computer or to the manual records for each colony. Some people like to record their assessments orally on a voice recorder and transfer data on return.

Alternatively, data can be recorded in the field directly onto a record card for each queen, and notes added as one works through the apiary. This can be useful as, by looking at the record card, we know the current position; is the queen marked; has she swarmed, etc.? Similarly, on return, these records can be fed into the computer.

At the end of each season the records can be summarised for each of our chosen characteristics, whether it be an average over the season, a total and a calculation (in the case of honey production) or the last recorded assessment (for swarming, for example) [see Chapter 2].

As with most aspects of bee improvement, record cards can be tailored to an individual's or a group's needs. Similarly a computerised version could be developed to work with one's own system of record-keeping. A group sharing the same record-keeping system may like to devise a system whereby different beekeepers could enter their data online to a shared system. A running summary of assessments could be automatically calculated. One, currently available, programme is the Bee Improvement Group (BIG) Stud Book which can be downloaded from the internet. (See Appendix)

The aim of good record-keeping is to allow colonies to be assessed over time so that an accurate selection of the most suitable stock for breeding can be made, whether for the rearing of queens or for drones.

 Bee Improvement Programme for Cornwall (BIPCo) Field Record Sheet

Site	Hive No.	Qu. Mk.	Date	Insp by	App	Te	Size	Sw	Brd/ He'th	Hon Yld	Comments/Needs

A Field record card for assessment of numerous colonies

Bee Improvement Programme for Cornwall (BIPCo) Record Card

Beekeeper		Site		Hive No.	

Queen Information

Date and Origin	Queen mark and description	Name of Strain	Type of hive or brood system

Date	Insp. by	Nat. App.	Tem per.	Size of col./fr of brood	Sw	Health Br.Pat	Fed	Hon Yld	Comments	Needs

A queen record card for recording performance throughout the season

4. Queen rearing methods

Regardless of how much progress we make in the breeding of bees, if queens are not reared well, they will never perform to their full potential. The best queens can only be reared by an abundance of nurse bees, well fed with honey, nectar or syrup and, perhaps most importantly, adequate stores of pollen. This allows the nurse bees to secrete the copious amounts of royal jelly that will get the larvae off to the best possible start.

A bewildering amount has been written about the numerous methods of queen rearing and it may not be helpful to add too much to this array of information as it can get difficult to know where to start. It is fine to try different methods but it is good to give each one a fair trial by repeating it several times before discarding it. Consistent results in queen rearing, particularly in our changeable climate, are notoriously difficult to achieve but by repeating a technique one can gradually iron out some of the variables and get better results. After a while one may have good reasons for trying another method; perhaps results are not as good as you wish or another method looks simpler and perhaps less stressful to the bees or the beekeeper.

Inducing the bees to produce queen cells is an artificial process, but within this constraint, one looks for a system which is, as much as possible, in harmony with the bees. Thus the bees are easy to manage, and adequate good quality queen cells can be produced in the simplest manner.

During the swarming season it is very easy to make increase from colonies producing swarm cells, but although this will produce well reared queens, it is a backward step in producing a bee with a low swarming impulse. Queens produced from the swarming impulse will, genetically, be more prone to swarming and in bee improvement this is something we should be aiming to move away from. By rearing our own queen cells, we have the opportunity to select and breed from queens which are less prone to swarming.

Use of 'starter' and 'finisher' colonies

The basic principles of queen rearing are that eggs or, more often, newly hatched larvae are presented to a colony or a part of a colony which is 'hopelessly queenless', that is, it has no queen, eggs or young larvae. Queen cells are reared under the 'emergency' impulse.

A similar effect can also be produced in a queenright colony by presenting larvae to a part of the colony that the queen is prevented from having access to, by the use of a queen excluder. This invokes the 'supersedure' instinct.

The colony which is presented with larvae for queen rearing is known as the 'starter colony' and usually, once the queen cells are underway, they are moved on to a 'finisher colony', or the colony may just be rearranged to convert it from a 'starter' to a 'finisher'. The question arises as to why we need the complication of a 'starter' and a 'finisher' colony. The first arrangement is, as the name implies, the best way to get the bees to commence building queen cells, whilst a 'finisher' colony is best able to provide adequate nourishment for the young larvae which are destined to become queens. Once the cells are sealed it just remains for them to be kept warm and safe from other queens, be it the old one or newly hatched ones.

One may already begin to see why there are so many methods of queen rearing; the setting up of 'starter' and 'finisher' colonies offers so much scope for variation. To add to this, there are different ways of presenting larvae to the 'starter' colony. Larvae (or eggs) can be presented in the worker comb that they have been laid in by the breeder queen. Alternatively, larvae can be transferred with the use of a Jenter or Cupkit, or they can be transferred by grafting, that is, the larvae are physically taken from the worker comb with the use of a grafting tool and placing in an artificial queen cup.

Transference of larvae using natural comb

A comb, laid up by the breeder queen can be used, as it is, or modified slightly, and presented to the starter colony as in the Miller Method, or the comb may be presented horizontally on top of a queenless colony in what is often known as the Hopkins Method.

Starter strip for Miller Method (J Phipps)

Board for holding brood frame horizontally above the brood nest for the production of queen cells using the Hopkins method

Transference of larvae using Jenter or Cupkit

The beekeeper may like to go a stage further and use a specialised piece of equipment in the form of a Jenter Kit or a Cupkit which consists of a small plastic cage holding approximately 100 plastic cells. This cage is fitted into a standard frame and usually sprayed with sugar syrup before placing in the brood nest for several days to 'acclimatize'. The breeder queen is placed in the cage for 24-36 hours. Often she does not start laying for some time so it pays to leave her in for a day and a half, say, cage on Day 1, pm and released Day 3, am, by which time she will have laid up the cells. Three days after release on Day 6, am, the eggs will have hatched and the young larvae are available for transfer by moving the plastic cells to the artificial queen cups on a cell raising frame in the starter colony.

Jenter frame inserted into brood comb (J Phipps)

Transferring larvae from Jenter kit to artificial queen cells (J Phipps)

Transference of larvae by grafting

Whilst the Jenter and Cupkit provide reliable methods of transferring larvae, they do require the purchase of specialist equipment. They also require considerably more preparation and planning than required for the grafting of larvae. With grafting it is possible to open the hive of a breeder queen and immediately take out young larvae and transfer to artificial queen cups in a starter colony. Grafting is a skill but one easily mastered with a bit of practice, particularly if you have a keen eye and a steady hand. The commonest method is to graft into JZBZ plastic cups which are readily available. They should be primed by placing the cell raising frame with plastic cups in the starter colony several days before they are required. The most popular grafting tools are thg Chinese ones which allow the larvae to be pushed off and deposited into the cell.

Chinese grafting tool

Cupkit and artificial queen cell frame for transfer of larvae

Unless you become particularly adept at grafting, on average the use of a Jenter or Cupkit will give you better returns in the form of completed queen cells as the larvae are totally free from interference. Also actual transfer of larvae tends to be quicker using kits than by grafting.

Queen rearing systems

Whole books have been written on the details of queen rearing methods. You may be fortunate to have someone experienced to show you how they rear their queens. Alternatively find a book or a method that sounds workable to you. I started with 'Queen Rearing Simplified' by Vince Cook which I followed implicitly and got good results. Very soon you will find you are modifying certain aspects to your own preferences, such as using plastic cups instead of making wax ones as he suggests.

A natural follow on from this is to use a Cloak Board which produces a very similar effect to Cook's method but with less manipulation of the colony. Both methods are, to some extent, slightly stressful to the colony, with periods of queenlessness and changes to the entrance positions, albeit for short periods of time.

For this reason, I decided to try another method as advocated in the paper, 'Rearing Queen Honey Bees in a Queenright Colony' by David Wilkinson and Mike A. Brown (see details below). No method is fool proof, particularly in our changeable climate, but this method produces good quality queen cells, and is less stressful to the queen rearing colony as it is kept queenright throughout. Some movement of combs between boxes is required but as the colony is never made queenless the bees are happier. The initial setup produces both a starter and finisher colony in one. Some failures are inevitable but this methos is as reliable as any other and produces many good quality queen cells.

The Queen Rearing Schedule using the Queenright System.

Many systems require a 'starter' and a 'finisher' colony but this system uses just one 'starter/finisher' colony. The colony to be used should be encouraged to build up on a double brood box (feeding as necessary) and is considered strong enough for queen rearing when bees cover the equivalent of about 20 National deep frames and include 8 to 12 frames of brood in all stages.

The empty graft frame, if using JZBZ cups, should be added to the hive for 'priming', a few days before it is required. The bees will build a circle of wax around the edge of the plastic cell cups.

On the day before grafting, or the receipt of Jenter cells, the colony is rearranged so that most of the sealed brood is in the top box and most of the unsealed brood and the queen are in the bottom below the queen excluder.. A comb of pollen should be selected and placed near to the proposed position of the graft frame. Also a frame of young larvae, preferably with some pollen stores, is placed adjacent to the graft frame.

If there is not a reliable nectar flow on, a stimulative feed of 1 or 2 litres can be given, in a frame or contact feeder. This light feeding can be given weekly if it is intended to keep the system running for the production of more than one batch of queen cells. Extra boxes can be added on top if more space is required to allow comb building to continue and discourage bees from filling in the spaces between the queen cells with comb (webbing). This webbing can lead to cells being neglected by the bees as well as making the distribution of queen cells difficult as they have to be cut out of the combs.

Larval transfer

On the next day the larvae from the chosen 'breeder' queen should be grafted into the JZBZ cell cups or, alternatively, the Jenter cups or Cupkit cells complete with larvae can be placed on the queen rearing frames. The graft frame with larvae can now be placed in the brood box immediately above the queen excluder next to the frame of unsealed larvae and pollen. When grafting, larvae should be protected from wind and sun and so working in a building or a vehicle can be helpful.

After 2 or 3 days, acceptance rates can be checked by lifting the frame out and observing how many cells have wax extensions and contain a larva on a bed of royal jelly. Sometimes emergency cells will be raised on some of the other frames of brood and these can be removed.

Acceptance levels vary according to many factors, the weather, the time of year, the condition of the rearing colony, food supply and so on. Greater consistency can be achieved through practice, getting to know the requirements, and by reducing the variables as much as possible. However, queen rearing, in Britain and Ireland will always, to some extent, be an erratic process and perseverance is required.

Distribution of Queen Cells

Grafting takes place when the larvae are on Day 4/5 of the queen rearing cycle. Queen cells can hatch on Day 15 or 16 so should be distributed to nucs for hatching and mating on Day 14. The day for distribution can be calculated by adding a week and three days to the grafting day; thus, if you graft, or transfer the larvae, on a Monday, distribution will be the following Monday plus three days, which is on the Thursday.

Queen cell (reared by Miller method) ready for distribution to mating nuc. (J Phipps)

Use of an incubator

Some people prefer to hatch the cells in an incubator in order to avoid the waste of setting up nucs with queen cells which, sometimes, do not hatch. A small poultry incubator set at 34 degrees C. is ideal. A high humidity level is required so the channels in the base of the incubator should be kept filled with water.

The queen cells need to be caged to prevent the hatched queens destroying the other cells and can be placed in the incubator any time after sealing but certainly by Day 14. Newly hatched queens will feed themselves in the cages if given a drop or two of 50:50 honey/water mix in the bottom of the cage and can be kept alive for several days in this way. It is, however, advisable to distribute the virgin queens to nucs as soon as possible after hatching for the welfare of the queen. These will be readily accepted in queenless nucs and can just be run into the entrance, particularly in the evening when all is quiet.

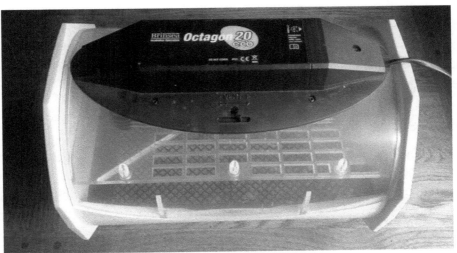

A poultry egg incubator used for hatching queen cells

Re-use of queen rearing colony

The queen rearing process can be continued by re-using the rearing colony, on a regular basis, if necessary. The colony should be re-arranged once a week; sealed brood frames from the lower box can be moved to the upper box, whilst frames in the upper box that have mostly hatched can be placed below the excluder. In other words it is a process of making more room for the queen to lay below whilst keeping plenty of brood (and therefore bees) above the excluder. A system of rotation, within the two brood boxes, can be developed in which the frames are moved round in a circle, putting them in at one end and taking out the other.

Within this method, a frame of young brood, with pollen if possible, is moved up, adjacent to the graft frame, or where the next graft frame will go. The queen does not have to be found for this manoeuvre, as the bees can be shaken or brushed off the combs before placing them in the upper box. In this way it is possible to rear one or two batches of queen cells per week.

Queen-rearing timetable

For full queen rearing timetable please see Appendix

Queen rearing records

When a programme of queen rearing is embarked upon it becomes imperative to keep accurate records. These would include a list of breeder queens and their current location, the dates of every procedure, which queen provided the larvae, the number of queen cells produced, which nucs and mini-nucs were provided with queen cells, numbers of laying queens produced, and so on. Without good records, one quickly loses track of all the details and monitoring the success rate at each stage of the process and the comparative qualities of the offspring of each breeder queen becomes impossible.

Conclusion

My preferred method of queen rearing is to use the queenright system and provide larvae by the use of a Cupkit.

Grafting is a skill worth mastering for the occasions when it is not feasible to use a Cupkit, such as if larvae are available for immediate use but the queen had not been placed in the Cupkit cage at the appropriate time.

5. Queen Mating - The use of nucs and mini-nucs

It may seem like a lot of work rearing queens but the biggest job of all is finding homes for all the queen cells or newly hatched queens. It is at this stage that the greatest risk of spreading disease occurs, as colonies get split up and distributed. If working in a Group, there may be movement of stocks, from more than one beekeeper, for making up nucs or for the mating of queens. Top priority must be given to health checks of the colonies (seek advice if necessary) to ensure that all stocks used are free from AFB and EFB. Attention, as always, should be paid to apiary and hive hygiene.

Most of us are familiar with the 'five-frame nucleus'. It is the common unit in which bees are sold; the reason being that a unit consisting of a young queen, three good frames of brood, plus stores, is self-sufficient and will expand rapidly, given adequate forage or feeding, and thus soon develop into a full production colony.

The use of nucs in queen rearing is slightly different as our aim is to provide a unit to allow the queens to mate rather than to produce a rapidly-developing production unit. Five-frame nucs are fine for the job, particularly if we wish to build up to a full colony quickly, or our aim is to have a nuc for sale at the earliest convenience. However, if we are aiming to produce queens for sale, or for our own use, there may be no rush to build up to strength and so we can opt to use smaller nucs or even purpose-built mini-nucs.

Smaller units require fewer resources and therefore a greater number can be produced from a given amount of stock. This can be important when faced with the prospect of finding homes for large numbers of queen cells or virgin queens. However this reduction in size comes at a price. Smaller units are more vulnerable and therefore require more nurturing. An eye must be kept on levels of stores and amount of space for bees and brood. One moment they may be on the verge of starvation and the next they may be outgrowing their home. Any lack of attention by the beekeeper may result in failure and this explains why the use of mini-nucs causes so much frustration. However, with regular attention to detail they provide a useful tool, producing mated queens with the use of very few bees, and in less time than it takes in a larger nucleus.

The Supply of Nucs

The supply of nucleus colonies, one for each queen cell or virgin queen, can be quite a challenge when a large batch of queen cells is maturing. There is a definite deadline (Day 14) when queen cells must be distributed. This can be extended slightly by caging the cells, either in the hive or in an incubator, but the preparation of nucs is a big commitment and takes a lot of work.

One method, which Vince Cook, (1986) describes fully, is to split the queen rearing colony into nucs. A colony on a double brood box, with the addition of two brood frames if using a National, would have 24 frames available, thus would provide material for 6 x four-frame nucs or 8 x three-frame ones. There should be enough bees in each nuc to look after whatever brood is provided, plus adequate stores. As long as they are queenless, and have no queen cells of their own, they will nurture the queen cell or virgin queen provided.

This is a simple method of providing the resources for a good number of nucs, especially if the queen rearing colony is not required to produce a further batch of queen cells. The rearing colony could be selected with this process in mind, by choosing a colony which, genetically, needs to be replaced with offspring from the chosen breeder queen. Perhaps the temper of this colony, which will later be broken into nucs, may not be as good as it should be, or the bees are not of the desired strain.

If one wishes to keep the queen rearing colony in tact to rear another batch of cells, other colonies, which have been earmarked for replacement and built up on double brood boxes, can be used for splitting into nucs when strong enough. This allows us to effectively cull out the worst of our colonies without wasting any resources. It is a good way of improving the quality of our stock as poor quality colonies are replaced with nucs headed by better quality queens from our selected breeder queens.

The Preparation of Nucs

Nucs can be prepared a day or two before the introduction of queens, or queen cells, and this helps to spread the work load. Also, if the nucs are well and truly queenless for a day or two, they will be keen to accept the queen or queen cell provided. It is worth checking such nucs for the raising of emergency queen cells. Although they may accept the one provided, they could already be developing other queen cells which can lead, even in small nucs, to premature swarming and the loss of a precious queen.

Multi-nuc box made from one brood box (J Phipps)

Single four-frame polystyrene nucs (J Phipps)

To make up nucs, simply select between 2 and 5 frames of bees with a mixture of brood and stores, probably varying from one of brood and one of stores in a 2 frame nuc to, say, 3 of brood and 2 of stores in a 5 frame nuc. When taking nucs from queenright colonies, it is advisable to locate and temporarily cage the queen first to avoid inadvertently putting her in the nuc. If the nuc is remaining on the same site a lot of the flying bees will return to the parent colony so extra bees will need to be shaken in. Too few bees will result in neglected brood. Best results are achieved by taking a newly made up nuc to a new site, thus avoiding the loss of flying bees resulting in over-weak nucs and, sometimes, setting up the risk of robbing.

Small nucs have few foragers so may rely heavily on the beekeeper to supply feed. They are particularly vulnerable in spells of bad weather and, later in the season, to robbing by bees and wasps. Entrances can be reduced to one bee space but once robbing has started it may be impossible to stop. Nurturing nucs through August and September can be particularly difficult so it may be worth aiming to get queen rearing finished earlier in the season.

Making up a drone-free nuc

If the nuc is to be transferred to a mating apiary so that the queen can mate with specific drones in the area, it is important to avoid any drones being transported with the nuc. The easiest way to make up a drone-free nuc is by the use of an additional brood box. Take the frames required for the nuc out of the parent colony and shake or brush the bees off the frames before placing in this second brood box. These frames should be replaced with drawn out comb or foundation in the original brood box, rearranging as necessary. A queen excluder is placed above the original brood box and the second box with the 5 selected combs, but no bees, placed above the excluder. For a 5-frame nuc this would be 2 frames of stores on the outside and 3 of brood in the middle. This can then be left for 2 or 3 hours or longer during which time the nurse bees will come up through the excluder to cover the brood. These drone-free combs, now complete with worker bees, can then be placed in a nuc box and transported to the mating site with no fear of contaminating the site with exotic drones.

The use of mini-nucs

Polystyrene mini-nucs of various designs are readily available. The Apidea is perhaps the most sophisticated but some prefer other designs which may have a greater capacity and are generally cheaper.

Whichever type is being used the best results are achieved by completely filling the food chamber with fondant (available in 12.5K boxes from many bakers).

The bees will need to draw out wax and so will consume a lot of stores to do this. Failure to fill the food chamber will allow the bees to occupy it and build wild comb in it. Frames, or top bars, should be prepared by attaching a strip of fresh foundation with melted beeswax. The mini-nucs are now ready for use.

One cupful (300ml) of bees is required for each mini-nuc. Bees can be collected from super frames which, being queenless and droneless, are easy to use. If bees are taken from brood frames, the queen must be avoided and the bees may need to be put through a queen excluder to filter out the drones. To collect the bees, it is usual to give them a light spray of water on the frame to reduce flying. They can then be shaken into a container (e.g. a round bowl) for scooping up by the cupful.

The entrances to the Apideas need to be closed before turning upside down and sliding the floor open for loading with bees. One 300ml cupful of bees can then be shaken into each mini-nuc, the box is tapped down, the floor pushed back into the closed position and the mini-nuc is returned to the upright position.

A ripe queen cell can then be inserted through the round opening in the clear plastic 'cover board'. Note that the frames in the Apideas need to be arranged differently according to whether using queen cells on the Jenter system or the JZBZ queen cups. For Jenter, two frames should be placed with their semi-circular notches in the top bars, facing each other to make a whole circle under the hole in the cover board. The JZBZ have smaller 'tops' so the frames can be arranged so that only one notch (a semi-circle) is under the hole in order to support the cell.

A similar process can be followed for other types of polystyrene mini-nucs. These are often less refined with no internal queen excluder and no cover boards. The capacity of these nucs is usually greater so more bees can be provided initially. Queen cells can be introduced by wedging between the combs.

The loaded mini-nucs can now be placed in a cool dark place for 3 days, with entrances remaining closed. This gives them time to start drawing the combs out and prevents the early absconding of the bees. During this time the queen will hatch and the mini-colony will start to function as a unit. By the third evening the mininucs can be placed in the mating apiary and the entrance opened so that, over the next few days, the queen can come and go on her mating flights. Once mated and laying, the queen excluder, at the entrance to the Apideas, can be slid across to prevent the colony absconding with the queen.

As soon as sealed brood has appeared in the mini-nuc, the queen can be assessed as to whether she is laying fertilised eggs, that is worker brood, and is, therefore, not a 'drone-layer'. The queen is now available for distribution and can be placed in a queen cage with workers for introduction to a nucleus colony. The mini-nuc can be used again after removal of the queen. After a short time-lapse, perhaps an hour or two, a new queen cell can be introduced and the process begins again. Because of the presence of brood, the colony is more firmly established and absconding tends to be less of a problem. Similarly any drone-laying queens can be replaced with a new queen cell.

Mini-nucs:
(a) Kieler mini-nuc

(b) and

(c) Payne's nuc box with feeder

(d) Apidea mating-nuc

6. Queen mating – The Role of the Drone

The importance of the drone in bee improvement cannot be over emphasised. The fact that queens and workers inherit 50% of their genes from the drones and the difficulty in controlling which drones mate with our queens is often cited as a reason why bee improvement is not possible for ordinary beekeepers. 'Good' drones can only be produced once we are able to select and breed from the best queens. It is imperative, in any bee improvement programme, to exert as much influence as possible over the input of the drones and thus increase the rate of progress that can be made.

The drones that a queen produces have the potential to alter the makeup of the gene pool in that area. Queens reared from a selected breeder will produce drones with genes inherited from that breeder, irrespective of what the new queens have mated with, as the drones are produced from unfertilised eggs. The process of repeatedly producing batches of new queens from selected breeders will improve the genetic quality of drones available in that area. If this is combined with the constant replacement of undesirable queens with these new queens, it can be seen how the beekeeper or group of beekeepers can have a positive effect.

Likewise, if a new queen is brought into an area, it is not just a matter of importing a queen in isolation, but new genes are being introduced into the area through the drones that she produces. An imported queen of a different sub-species adds genes to the gene pool from bees which have no local adaptation and no history of good performance in the area. Importation of different sub-species, that is of bees which have been adapting to their own geographic climate for thousands of years, is the biggest obstacle to the development of a local strain which can be selected and improved. It quickly negates the benefits of any natural and artificial selection carried out in a bee improvement programme.

Drones on alighting board in the warmer part of the day. However, native drones will fly at cooler temperatures than other sub-species. (J Phipps)

Options Available for the Mating of Queens

Various theories and methods have been suggested for achieving matings of queens with the favoured drones, such as early or late season queen rearing, restricting queens and drones to later day-time flying and even controlling light and temperature to trick drones into responding as if it is the right time of day for mating. An easily achievable and reliable method has, probably, yet to be found, leaving three options for achieving total or partial control of mating:

1. Instrumental insemination
2. The use of isolated mating apiaries

3. The drone flooding of an area

Instrumental insemination (I.I.) is the hi-tech route, requiring expensive equipment and expertise. It allows complete control over the mating process taking selection to a new level as the beekeeper is able to choose the male and female lines. I.I. is often regarded as the answer to all the imperfections of using open matings but, whilst some operators and breeding programmes do use it successfully, it is fair to say that few beekeepers master the technique on a small scale. Even if the process can be operated successfully, it does not come without its problems; there is a danger of narrowing the gene pool so much that inbreeding occurs, something we are unlikely to see with natural matings. Queens mated by I.I. are generally only used for producing more breeding stock rather than for heading honey production colonies.

The emphasis in this book is that bee improvement can be practically applied and achieved by all beekeepers. Whilst instrumental insemination is a technique that some breeders consider worth mastering, it is generally a step too far for the average beekeeper and, it seems, one which will not generally be of benefit in small bee improvement programmes.

The second option is the use of isolated mating apiaries. This plays a very important role in many breeding programmes around the world. For example, numerous islands off Denmark and the north coast of Germany are used by beekeepers for this purpose. However, a lack of convenient sites in the British Isles makes this a difficult and expensive option to take advantage of. Total isolation on the mainland is not a feasible option for most of us, bearing in mind that the range of drones may be 10km or 6 miles and that drones are not fixed to one particular hive but will be accepted in any colony.

Whilst total isolation is difficult, a degree of geographical isolation can often be achieved. A peninsula, largely surrounded by sea, or a valley in an area of high land generally unsuitable for beekeeping, may help to reduce the number of undesirable drones reaching our newly reared queens. The use of local topography and natural microclimates can assist in selecting good mating sites. This partial isolation combined with the third option, that of drone-flooding of an area, can help to increase the number of desirable matings.

Drone flooding is the option that most of us are left with, that is, the saturation of an area with the preferred drones. Ideally we need to have control over most of the colonies in an area and so co-operation with other beekeepers is usually essential. By repeatedly producing and distributing queens and encouraging our best queens to produce more drones through the use of drone foundation and stimulative feeding we can influence the drone population in an area.

Some areas, perhaps with high levels of imports, will be notoriously difficult to influence but we should persevere with our principles and improvements are possible. By repeatedly replacing the worst queens with those of the right strain from our breeder queens the chances of achieving better, less hybridised matings are gradually increased. Very few of us are in the perfect position but if we do nothing then things will never improve. Relying on the use of 'drone-flooding' may seem like an imperfect system, especially as we are at the mercy of beekeepers in our area who may continue to use imported bees, either from choice, or from ignorance of an alternative, or because of the unavailability of an alternative source of bees.

This has been an increasing problem, over the last few years, with the growing numbers of new beekeepers who tend to be supplied with nucs headed by imported queens. It is in the interests of local improvement groups to provide an alternative for neighbouring beekeepers, but it takes time to reach a position where this is possible. Despite the imperfections of 'drone-flooding' there are some factors which may work in our favour.

Mating Advantages of Native Drones

In warm fine weather (over 20 degrees Celsius) drones congregate in large groups known as drone congregation areas (DCAs). The competition amongst the drones in these areas to mate with queens is intense. In continental Europe, we are told that this is the only mechanism for the mating of queens. In some seasons in many parts of Britain and Ireland temperatures scarcely rise above 20 degrees Celsius and yet queens still get mated. In cool summers in areas with high proportions of bees of exotic strains we hear tales of woe about the poor mating of queens. It seems likely that the native strain can produce viable matings even in spells of poor weather. This is known as 'apiary vicinity mating' and appears to be of great importance to the native bee. It is probable that the native drone has an advantage in our cool changeable climate perhaps being a stronger flyer in poor weather.

Many beekeepers notice how drones congregate in hives and nucs with the presence of virgin queens. One can assume that if the virgin queen leaves the hive on a mating flight, perhaps in a break in the weather, many drones would follow. It seems that bees adapted to our maritime climate work differently to those adapted to the warm settled summers of continental or Mediterranean climates which allow the formation of DCAs.

There may be other mechanisms which encourage bees to mate within their sub-species that we still know little about; one theory is that different sub-species prefer to mate at different heights; another is that pheromones vary between sub-species, thus making attraction within single sub-species more likely. Perhaps this is just wishful thinking, but there must be reasons why some colonies appear to have survived as relatively pure examples of A.m.m. with no assistance from beekeepers and despite the presence of other sub-species in the area.

Establishing a strain through drone-flooding

For bee improvement we need as many desirable drones as possible in the area. The more queens of the native strain, regardless of what they have mated with, the better are the chances of our new queens mating within the strain. Thus if last season's queens were produced from good breeder queens, though their offspring may be disappointingly hybridised, we at least have the consolation of knowing they are producing good drones, which will help with the current season's matings. This is one of the keys to bee improvement which can really help us to bring up the quality of bees in an area. It depends on repeatedly producing new generations of queens from good breeders. In this way we are able to put out 'pure-bred' drones which will help to improve the matings of our queens.

If the drones are not of the right strain, though, this can set back our improvement programme, as hybridisation, once again, prevents us from achieving a consistent quality in our colonies. To establish a native strain in an area which may previously have been dominated by hybridised bees, we start by rearing queens from our chosen breeder queen. Although the new queens will mate with the local drones producing hybridised colonies, the drones produced by these young queens will only have inherited their chromosomes from their mother and so all their genes will be from our breeder queen. They are therefore of our chosen strain and if we produce enough queens, the area will become flooded with drones of our selected strain in the next season. The queens we rear now will have a good chance of mating with drones of the same strain, allowing us to develop our native/near-native strain. By repeating the process and perhaps increasing the number of colonies in an area our chances of establishing the strain will increase.

On the whole, the bigger, more prolific colonies will produce the most drones. There may be exceptions; a good honey producing colony may not produce as many; a drone-laying queen may produce more, but this tends to be a short-term phenomenon as the colony quickly weakens, and the drones will not be the strongest, having been reared in worker cells. If we are fortunate enough to dominate the area with native drones, nature will help to select the best, as the drones which successfully mate with our queens will come from the most vigorous colonies in the area, and so are contributing to genetic improvement.

Drone production

It is estimated that a colony can produce 20,000 drones during a season, though all maturing at different times and it is thought that this number can be more than doubled by stimulative feeding and the provision of drone comb or foundation.

Drones are sexually mature when 12-14 days old, so from egg to mature drone takes about 38 days (i.e. 24 + 14) whereas the queen can mate from about 5 days old, so from egg to maturity takes 21 days (16+5). This is a difference in maturity times of 17 days so it is usually recommended that drone producing colonies be stimulated at least 3 weeks before 'grafting day' for the queen production, particularly for early season queen production.

It is reassuring to know that even if the colonies produced from our new queens do not appear to have bred true, they will still be producing 'native' drones, if produced from a good breeder queen. The next generation of queens from these colonies, though, perhaps produced from swarming, will be hybrids so these colonies should ultimately be earmarked for breaking up into nucs for newly reared queens. Thus they will be prevented from releasing hybrid queens into the area (through swarming).

The aim must be to keep extending the mating area and also, if possible, to establish other similar areas. However, it does require a lot of colonies, and the production of many queens, to produce the desired drone population. The establishment and maintenance of a good mating area requires renewed effort every season and, either, the removal or break-up of hybridised colonies for nuc production.

7. Working as a Group

Beekeepers working as a Group can pool skills and resources. A group can manage a larger number of colonies than an individual, allowing the possibility of achieving a more reliable mating area. Equipment, effort, responsibilities and enthusiasm can be shared, meaning that jobs can get done at the correct time, even when some members are unavailable.

Beekeepers are often individuals who prefer to work on their own; indeed this may be one of the attractions of beekeeping. However, even those working on their own may find advantages io linking up with other individuals for certain tasks, such as the exchange of breeding material or for the use of a shared site for the mating of nucs.

When setting up a Group, it is important to gather together beekeepers who share common aims and a determination to continue working towards the goals no matter what problems occur on the way. Setbacks are inevitable and the group will be most fragile in the first 2 or 3 seasons when there may be little to show for all the effort put in. Aim to make some progress each season, even if it is only increased experience which will help to achieve better results in following seasons.

It is good to talk to others about bee improvement and how it may be achieved but do not expect everyone to agree with what may seem obvious to you. Progress will be more easrly achieved by working with a few like-minded people rather than trying to include those that are not fully committed, or do not understand the basic principles. By working quietly with those who show a genuine interest, the results will eventually speak for themselves. Lack of beekeeping experience should not exclude people as their skills will develop rapidly when repeatedly involved in queen rearing.

It is fine to start as an informal group which works well in many situations. A more formal approach only becomes necessary if the Group wishes to raise funds and then a formal constitution will need to be adopted and a bank account opened. Funding, of course, has advantages but brings increased responsibility and bureaucracy. The Group becomes able to purchase valuable equipment which should be clearly labelled as assets of the grouy, perhaps by branding. Assets have to be accounted for and records kept of their whereabouts. Membership fees may be introduced if there are definite benefits of membership, and by joining, members are able to show their commitment to the group.

For further information on the setting up and running of Bee Improvement Groups see:

The Bee Improvement Group Handbook, a BIBBA Publication

PART 3:
SUMMARY OF 'THE PRINCIPLES OF BEE IMPROVEMENT'

The Principles of Bee Improvement

1. **The relationship between humans and honey bees is even more important today than it has been in the past. It is crucial to world food production and to the welfare of the wider environment.**

    There has been a long association, going back into pre-history, between human beings, *Homo sapiens,* and honey bees, *Apis mellifera.* As agriculture has developed and world populations have expanded the importance of the honey bee to food production and to the wider environment, through pollination, has increased. It is essential that this long association is maintained in a sustainable way. Custodians of the countryside can contribute to this sustainability by providing an environment safe to pollinators and rich in available forage.

2. **Beekeepers are essential to the provision of an adequate bee population and in order for this to be a sustainable relationship they require an adequate return on their investment.**

    The beekeeper plays an important role in allowing the honey bee to provide a 'free' pollination service. Whilst large-scale beekeepers may be able to charge a fee for this service the small-scale beekeeper gets no such recompense. In a rich society, with adequate leisure time, this may be provided by altruistic beekeepers or those for whom the pleasure of the craft and some small returns are enough. Long-term sustainability, though, is reliant on beekeepers achieving an adequate return on their investment, usually in the form of honey production.

3. **'Bee improvement' can increase the beekeeper's return and, thus, is important to the long-term sustainability of beekeeping.**

    An increase in the productivity of the honey bee offers a greater return for the beekeeper and thus is a factor in the long-term sustainability of beekeeping. In a densely populated world, qualities such as non-aggression and low-swarming become more important, and can only be achieved through 'bee improvement'.

4. 'Bee improvement' should be based, firstly, on 'natural selection', that is, on the principle of the 'survival of the fittest'.

Nature maintains healthy populations, adapted to the local environment, through 'the survival of the fittest', and beekeepers should work with the same principle. In nature the fittest stocks are the ones that survive and breed. In beekeeping, we want robust bees that over-winter well and are not prone to disease. This can be achieved by the avoidance of bringing in unproven stock from another area or from abroad. The import of stock from abroad brings in new genes, generally of other sub-species, which have not been tested in the local environment.

5. 'Bee improvement should be based, secondly, on 'artificial selection', that is on the qualities which are required by the beekeeper.

A bee which is purely selected by nature may be of little use to the beekeeper. The beekeeper can build on the advantages bestowed by 'natural selection' by further selecting for other desirable qualities, such as 'good temper' and 'productivity'.

6. Breeding stock should be selected through the careful assessment of colonies combined with an accurate system of record-keeping.

A system of record-keeping should be used which is simple and objective, so that any operator can assess or mark colonies to the same standards. This will allow different colonies, in different apiaries, perhaps owned by different beekeepers, to be rationally compared.

7. In order to reduce variability within offspring, bees should be selected within a strain, that is, within a more uniform population. In areas where the honey bee is native, the native sub-species for that area is likely to be the dominant strain and thus the easiest one to use in order to make progress in bee improvement.

The sub-species of honey bees have become very mixed in most parts of the world. It is essential in 'bee improvement' that bees breed 'true to type', that is, offspring resemble their parents rather than producing offspring of random variability. In most areas it is doubtful whether it is possible to ever return to a totally pure sub-species but the aim should be to at least to work within a 'near-native' strain.

8. **Improvement in the quality of bees depends on a system of the selection of the best and the culling of the worst.**

Improvement of stock is achieved by selecting the best and getting rid of the worst. The more beekeepers in an area working towards the same goals, the more chance there is of achieving good results.

Resources do not have to be wasted; stocks of bees which are not of the required strain, or have undesirable qualities such as bad temper, can supply the bees for nucleus colonies which can then be provided with home bred queens of the required strain.

A rigorous system of breaking down undesirable colonies into nucs and re-queening will, over time, reduce the number of undesirable drones being produced in the local environment.

9. **A strain can only be developed and maintained in an area if due regard is paid to the importance of the drone.**

The drone is crucial to the maintenance of a strain. Whatever mating areas are available, success ultimately depends on dominating the area with drones of the selected strain. This is achieved through producing numerous queens of the right strain; regardless of what they mate with, they will produce 'good' drones to mate with the next generation of queens.

10. **Bee Improvement can be achieved from a 'top-down' approach, that is by starting with 'pure' stock, or from a 'bottom-up' approach, that is, by selecting and refining the stock that is available in an area, or by a combination of the two.**

The 'top-down' approach has the advantage that a lot of the work involved in the development of a strain has already been done but it may prove difficult to maintain that strain without further 'imports' of 'pure' stock.

The 'bottom up' approach may require a lot of work in developing and refining the strain but it has the advantage of producing stock with local adaptation and of developing systems in which the strain is strengthened.

CONCLUSION

The honey bee has survived in the British Isles since the retreat of the last Ice Age. Despite rumours of its demise, it is unlikely to disappear and will adapt to any environmental changes which occur. The importance of the honey bee to mankind has increased as world populations have grown and the dependence on pollinators for world food supplies have become more important.

The health of the honey bee has been greatly affected by the parasitic mite, Varroa. In time the honey bee will adapt and be able tolerate Varroa, whether through natural selection or through the efforts of beekeepers in selecting a bee more able to cope. This process is likely to be speeded up by the use of hardy stock which is not constantly undermined by the importation of untested stock. It stands to reason that winter losses are likely to be higher if we choose to use bees which are not adapted to local climatic and environmental conditions.

Progress in improving the quality of our bees has been virtually non-existent as we have relied on others to produce good bees for us. The standard 'top-down' approach in which high quality queens are produced and made available for beekeepers to buy has not produced any improvement in standards. Now is the time to try an alternative approach, that of ordinary beekeepers making the most of what they have got, and building up from the grass roots. Even though this inevitably means starting from a low base in terms of the quality of our bees, this may be the route that can actually produce continuous progress.

The current interest in bee improvement and the realisation that 'home-grown' bees may well be better than 'imported' ones bodes well for the chances of raising the standards of our bees. By producing bees which are tuned to the vagaries of our weather and the local environment, we are more likely to see our bees survive and perform well.

Ordinary beekeepers can prove that progress in improving the quality of bees can be made through working together in a common aim. It is a huge challenge, but one that has never been more necessary than now.

APPENDIX 1:

Queen rearing timetable using the Cup-kit or Jenter system with the 'Queen-right method

Appendix 1:

Queen rearing timetable using Cupkit and Queenright Method

Day from laying	Action on breeder queen colony	Action on queen rearing colony	Example day
-7 to -21		Feed as necessary to build up the queen-rearing colony in a double brood box	
- 3	Spray Cupkit cells in cage with weak sugar syrup and place frame with Cupkit in centre of brood nest of 'breeder queen', to 'prime'		Wednesday
-1 (afternoon)	Place breeder queen in Cupkit cage to lay up cells for transfer to queen rearing colony		Friday
0	Laying day (no action)		Saturday
+1 (morning)	Release queen from cage (checking cells are laid up with eggs)		Sunday
+3		Set up queen rearing colony to receive the larvae i.e. queen below queen excluder in bottom brood box; frames arranged in top brood box: - 1.Stores 2.Open stores 3.Pollen 4. (position for queen rearing frame) 5.Unsealed brood 6.Sealed brood, etc.	Tuesday
+4 (morning)	Larval transfer day: transfer cups with day-old larvae to queen rearing colony		Wednesday
+14 (morning) (1 week & 3 days after larval transfer day)		Transfer queen cells to nucs or mini-nucs, or cage and place in an incubator	Saturday
+15/16	Queen cells hatch		Sun/Mon
3 to 4 weeks later	Queen can be checked for sealed worker brood		

N.B. If grafting, will need to follow instructions for 'queen rearing colony' and graft larvae on Day 4. The grafting frame with JZBZ cups should be placed in rearing colony (to prime) 2 or 3 days before.

APPENDIX 2:

To find downloads of DrawWing and MorphPlot for bee wing morphometry, and the Bee Improvement Group (BIG) Stud Book for record keeping, follow the link below to the Stratford-upon-Avon BKA:

http://www.stratfordbeekeepers.org.uk/Links.htm#Morphometry
A link to a download for Beemorph can be found at:
http://www.hockerley.plus.com/
An example of a BIPCo Queen Record Card can be found on
http://www.bipco.co.uk/

REFERENCES:

Atkinson, John H., (1999), Background to Bee Breeding

BIBBA, (2014), The Bee Improvement Group Handbook

BIBBA, 1990) Guidelines for Bee Breeding

Brother Adam, (1980), Breeding the Honeybee

Brother Adam, (1983), In Search of the Best Strains of Bee

Brother Adam, (1987), Beekeeping at Buckfast Abbey

Cooper, Beowulf, (1986), The Honeybees of the British Isles

Cook, Vince, (1986), Queen Rearing Simplified

Dews, John, and Milner, Eric, (1991), Breeding Better Bees

Holm, Eigel, (2010), Queen Breeding and Genetics

Pritchard, Dorian, (2011), Inbreeding Pt 1 & 2, BIBBA?Bee Improve. 35 & 36

Robson, W.S., (2013), Reflections on Beekeeping

Ruttner, Friedrich, (1988), Breeding Techniques and Selection for Breeding of the Honeybee

Ruttner, F., Milner, E, Dews, J., (1990), The Dark European Honey Bee

Wilkinson, David and **Brown**, Mike, (2002), Rearing Queen Honey Bees in a Queenright colony, Central Science Laboratory

Woodward, David, (2007), Queen Bee: Biology, Rearing and Breeding

ACKNOWLEDGEMENTS

My understanding of 'Bee Improvement' has developed through the works and inspiration of BIBBA, the Bee Improvement and Bee Breeders' Association.

It has been possible to test theories and methods and discover what works best, by working with, and receiving support from the beekeepers I work with and fellow members of BIPCo (The Bee Improvement Programme for Cornwall.

Our progress, as a Group, has been assisted by funding provided by East Cornwall Local Action Group, Rural Development Agency, which allowed us to provide equipment for queen rearing at several apiaries in East Cornwall. Benefits to local beekeepers have included opportunities to improve their skills and to participate in a programme to improve the productivity and temper of locar bees.

BIPCO was formed in 2009 by a small group of beekeepers who shared the common aim of wanting to improve the quality of their bees.

INDEX

appearance of native bee	34, 38-42
artificial selection	9-10
assessment of stock	23
assessment of stock, behaviour	24-30
assessment of stock, within a strain	30-42
assessment, record keeping	41-42,43-44
biosecurity	2
bee breeding/bee improvement	10
bee improvement/bee breeding	10
bee improvement, principles,summary	69-72
breeding true to type	10-12
brood pattern, assessment	28-29
characteristics, selecting	25
compatibility of sub-species	13-16
diversity within native strain	20-21
drone, role of	62-63
drone flooding	63-67
drone production	67
environmental factors, re assessments	24
group, working as a	67-68
hereditary factors, re assessments	24
health, assessment	28-29
hybrid vigour	12-13
hybridisation	16-18, 30-31
importing queens	6-8
inbreeding	19
mating, of queen	55
mating options	63-65
mating, use of nucs	55-58
mini-nucs, use of	58-63
morphometry	34-37
native bee, use of	19-20
native drones, mating advantage	65-66
native strain, identification	38-40
natural selection	9
nucs, use for mating	55-58

principles of bee improvement	69-72
productivity, assessment	29-30
pure strain, development of	31-32
purity of strain	21-22
queen rearing	45-54
record keeping	43-44
selection within a strain	18-19
swarming, low propensity, assessment	26-28
temper, assessment	25-26
principles of bee improvement	74-77
productivity, assessment	31-32
pure strain, development of	34-35
purity of strain	23
queen rearing	48-59
record keeping	46-47
selection within a strain	19-20
swarming, low propensity, assessment	28-30
temper, assessment	27-28

Lightning Source UK Ltd.
Milton Keynes UK
UKHW051543110221
378614UK00002BA/8

9 781908 904621